D1539079

Monitoring Land Supply with Geographic Information Systems

Monitoring Land Supply with Geographic Information Systems

Theory, Practice, and Parcel-Based Approaches

**Edited by
Anne Vernez Moudon**
University of Washington, Seattle

and Michael Hubner
Consultant to the Suburban Cities
Association of King County, Washington

John Wiley & Sons, Inc.
NEW YORK / CHICHESTER / WEINHEIM / BRISBANE / SINGAPORE / TORONTO

ISBN 0-471-37163-7

10 9 8 7 6 5 4 3 2 1

To Louisa and Cota

Anne Vernez Moudon

*To all the members of my family, especially to the memory of my
grandfather, Henry Bellinger, and to
Miriam Hirschstein for her support and affection*

Michael Hubner

Contents

Acknowledgments

This book were made possible by the generous support of the Lincoln Institute of Land Policy, which has long championed research and education on land monitoring. We especially want to thank Roz Greenstein, Director of the Institute Program in Land Markets, who played a dual role as infallible advisor and enthusiastic advocate of the project. Her help in setting the research program as an integral part of the Lincoln Institute's educational agenda was invaluable. She was also instrumental in turning the project into a book.

The study entailed extensive review of the literature on land monitoring and GIS as well as interviews with numerous scholars and practitioners in these fields. We are very grateful to those who generously shared their knowledge and experience with us during the course of this project; many of the people contacted during the course of the research are listed in Appendixes B and C. The project also depended on early inputs from an advisory board, which included Nicholas Chrisman, Bob Filley, Gary Pivo, and Paul Waddell of the University of Washington; Nancy Tosta of the Puget Sound Regional Council; and Carol Bason and Tom Nolan of the City of Seattle. Sharing facilities and equipment with the Center for Community Development and Real Estate at the University of Washington for the duration of the project brought an untold number of benefits.

Some of the chapters originated as papers presented at a seminar, "Parcel-Based GIS for Land-Supply and Capacity Monitoring," at the University of Washington in May 1998. Lively exchanges during the course of the seminar led to substantial revisions of the early drafts as well as to additional papers covering the new issues and topics raised. The seminar discussions greatly benefited from the input of practiced professionals involved in land monitoring at Portland Metro and at various agencies and jurisdictions of the central Puget Sound region. We would like to thank all the participants for their rich contributions and enthusiastic support of the project—a list of seminar participants is provided in Appendix D. We also wish to thank the staff of the College of Architecture and Urban Planning and the Department of Urban Design and Planning at the University of Washington for their help in organizing the seminar, especially Sandy Houser and Dorothy Sjaastad.

The book also gleans much material from the fruits of Michael Hubner's nine months of independent research, which culminated in a master's thesis in June of 1999.

Louisa Seferis helped transform the early research report into book chapters by expertly entering the numerous revisions made.

From the outset, we conceived of this work as an effort to bring an "user's" critical perspective to bear on the technical realm of GIS in planning. Our own backgrounds in the study of urbanization and development processes, the evolution of urban form, and land use and growth management planning methods, informed our research on urban land supply monitoring. However, the process was not painless, especially in the need to quickly catch up with several decades of GIS development and land monitoring practice. Our express gratitude to all of the people and organizations that freely lent their support does not take away from our final responsibility for this book. We hope that this work fulfills some of our original intent, namely to help planners improve their understanding of how cities are made, and thereby improve the quality of those cities in the future.

ANNE VERNEZ MOUDON AND MICHAEL HUBNER

Seattle, July 1999

Contributors

PROF. MARINA ALBERTI, Department of Urban Design and Planning, University of Washington, Seattle

PROF. WILLIAM BEYERS, Department of Geography, University of Washington, Seattle

PROF. SCOTT A. BOLLENS, Department of Urban and Regional Planning, University of California at Irvine

PROF. KENNETH J. DUEKER, Center for Urban Studies, Portland State University, Portland, Oregon

MR. MILES ERICKSON, Center for Community Development and Real Estate, College of Architecture and Urban Planning, University of Washington, Seattle

PROF. DAVID R. GODSCHALK, Department of City and Regional Planning, University of North Carolina at Chapel Hill

MCSHANE HOPE, Washington State Department of Community, Trade, and Economic Development, Olympia, Washington

PROF. LEWIS D. HOPKINS, Department of Urban and Regional Planning, University of Illinois at Urbana-Champaign

PROF. GERRIT J. KNAAP, Department of Urban and Regional Planning, University of Illinois at Urbana-Champaign

PROF. ZORICA NEDOVIC-BUDIC, Department of Urban and Regional Planning, University of Illinois at Urbana-Champaign

Ms. LORI PECKOL, Planning and Community Development Department, City of Redmond, Washington

PROF. GEORGE ROLFE,   Department of Urban Design and Planning, University of Washington, Seattle

MS. NANCY TOSTA,   Tosta Enterprises, Seattle, Washington

PROF. RIC VRANA,   Department of Geography, Portland State University, Portland, Oregon

PROF. PAUL WADDELL,   Department of Urban Design and Planning, University of Washington, Seattle

PROF. FRANK WESTERLUND,   Department of Urban Design and Planning, University of Washington, Seattle

Introduction

Anne Vernez Moudon and Michael Hubner

This book documents the use of geographic information systems (GIS) in monitoring urban land supply. Monitoring land is a well-established activity of many local governments in the United States. Recent developments in parcel-based GIS are now opening up new opportunities. This work explores how these new GIS can serve the purposes of current land monitoring efforts and how they may change them.

A first focus for this book is land monitoring in urbanized regions, as opposed to areas in primarily rural or agricultural use and wilderness areas. Urbanized regions differ from other land areas in that the use of land—both actual and potential—is defined by a complex web of market forces and public policies. The use of land in urbanized regions is highly dependent on the level at which the land is serviced with infrastructure, including electricity, water, sewers, communications and transportation systems, and educational and recreational facilities. Urban land is also characterized by comparatively small spatial units of ownership and a high rate of change in ownership and use. In addition, it is subject to elaborate land use regulations and land management practices (Kivell 1993; Kaiser et al. 1995).

Concern about the continued spread of cities in the United States and the growing inability to cover the associated costs of providing urban infrastructure have raised the profile of land monitoring within urban and regional planning. At issue are both the available *supply* of land and the *capacity* of that land to accommodate future development. Land supply and capacity monitoring— LSCM for short—comprises the multiple activities involved in measuring, analyzing, and evaluating supply and capacity over time.

Monitoring Land Supply with Geographic Information Systems, edited by Anne Vernez Moudon and Michael Hubner ISBN 0 471371673 © 2000 John Wiley & Sons, Inc.

The practice of LSCM has required the use of increasingly sophisticated land information systems (Dale and McLaughlin 1988). In turn, the use of these systems has allowed planners to reach beyond a traditional focus on land use and zoning and to consider causal relationships between land utilization, land and infrastructure development, fiscal impacts, and market forces (Bollens and Godschalk 1987; Jacobs 1988).

The tax-lot parcel is an increasingly essential building block of land information systems. Previous efforts to describe a preferred model for collecting land data have emphasized the parcel as a central element, suggesting specifically the maintenance of a "parcel file" for land supply monitoring (Godschalk et al. 1986) or referring to a more generic "multipurpose cadastre" (Dueker and Kjerne 1989; Dueker and DeLacey 1990). Because it is the smallest unit of geography that can be managed and controlled by one person or entity, the parcel continues to be considered an important unit of data collection and analysis, which must be considered along with other levels of geography (Enger 1992; Bollens 1998).

Parcel-based geographic information systems—or PBGIS—are a second focus of this book. These relatively new systems are defined by the inclusion of parcels as a primary spatial layer. The digitization of parcel boundaries allows parcels to be represented and analyzed as discrete areas in a PBGIS. Tabular data files containing land records and assessor's files can thereby be linked to individual parcels via a unique identifier such as the parcel identification number (PIN). In addition, overlaying parcels with nonparcel spatial layers allows many variables, such as soil conditions and transportation infrastructure, to be considered precisely in relation to spatial patterns of land ownership, regulation, and development. PBGIS therefore provide a means for representing and analyzing various data about land at a higher level of spatial resolution than achieved with previous applications of GIS based on relatively coarse geographies, such as analysis zones, planning areas, or generalized land use polygons. The parcel level of resolution also corresponds closely to real-world transactional activity related to urban land.

THE PAST TEN YEARS

The state of the art in land supply monitoring and land information systems was last reviewed more than ten years ago (Godschalk et al. 1986). Since then, technological advances have continued at a rapid pace, highlighted by an increasingly comprehensive and widespread integration of GIS into the Automated Land Supply Information System described by these authors. At the same time, however, these advances are only slowly being met with the changes in the organizational structures that are required to apply the new information technology to planning analysis. Indeed, the current state of practice reflects the findings of the aforementioned authors. Although, on one hand, "computerized land supply tracking systems increase the comprehensiveness and currency of both public and private land-related information," technical capabilities appear "more advanced than the institutional capacity of governments to use land information systems effectively" (Bollens and Godschalk 1987, 315).

Three primary areas of change have affected the state of practice in land use monitoring and capacity analysis over the past decade. One is the extraordinarily rapid geographic spread of urban areas relative to increases in population. As urban sprawl, the conurbation of historically separate urban centers, the emergence of new suburban centers, and leapfrog development patterns have continued apace, many cities have evolved into city-regions. Land planning and, consequently, land monitoring have become regional in scope, involving numerous jurisdictions and multiple levels of government. The size and complexity of these urban areas pose significant challenges to developing appropriately detailed regional land supply databases (Kivell 1993; Downs 1994).

A second area of change is the increased level of concern for and awareness of how patterns of contemporary urbanization affect both natural and human-made systems. These concerns have spurred a multitude of legislative actions, both to alter urbanization patterns and to mitigate their impacts through a variety of mechanisms. It has become common for governments to encourage dense and compact urban development via urban containment strategies, ordinances to ensure the provision of adequate public facilities, farmland and open space preservation programs, shoreline and watershed management programs, and so on. Cast under the label of growth management or "smart growth," the new orientation in land use planning reflects a recognition that not only is urban land a finite resource, but that current land development patterns impose costs that local governments are increasingly unable, or unwilling, to bear (De Grove 1992; Peirce 1997; American Planning Association 1998). In this context, monitoring land and its development has become an integral, and often legally mandated, aspect of many jurisdictions' managerial responsibilities.

A third area of change relates to significant advances made in information technology. The rapid evolution of GIS technology has been marked by an exponential increase in the availability of spatial and spatially referenced data, as well as by increased possibilities to integrate a multiplicity of databases (Chrisman 1997; Klosterman 1997). In particular, the profusion of parcel-level data now enables planners to analyze in great detail the locational and spatial characteristics of a land area, along with its social, economic, and environmental dimensions. Improvements in both hardware and software have increased capacities for data storage, processing, and manipulation, allowing land monitoring programs to collect and manage land-related data at high levels of detail over large metropolitan regions.

CASE STUDIES OF "BEST" PRACTICES

Technical advances and policy reorientations notwithstanding, this book projects a context for land monitoring that is in advance of practice. A review of the literature on GIS and land monitoring, complemented by numerous phone interviews with people involved in both fields, indicates that, with major exceptions, relatively few jurisdictions in the country have used PBGIS for land monitoring and capacity analysis. At the same time, however, it is estimated that 25 percent or more of jurisdictions across the country are in the process of building a PBGIS, including most cities with populations of more than 100,000 (Crane 1997; Huxhold 1997; Kollin et al. 1998). As parcel-based systems become more widely available, planners will increasingly utilize these data, both for daily tasks and for long-range monitoring and analyses related to land supply.

Given the anticipation of widespread use of PBGIS and the significant changes they are projected to make in methods of land use monitoring, this book highlights three case studies to illustrate "best" practices and advanced applications of PBGIS for land supply and capacity monitoring. First, Portland Metro, Oregon, has a 20-year history of growth management, an established legal and institutional apparatus to monitor land supply, and a sophisticated information system to manage its lands. This multijurisdictional effort offers a well-engineered and tested set of tools and methods, which include the use of mixed spatial units (including parcels), the integration of raster and vector GIS, the periodic application of aerial image data, and ongoing monitoring of regional "Performance Measures" related to land use, land development, and housing.

Second, Montgomery County, Maryland, has a distinguished history of land planning and a tested system of releasing land for development in accordance with infrastructure delivery. Although the county has historically used a variety of non–parcel-level and non-GIS computerized land monitoring tools, it is now in the process of incorporating PBGIS analysis in support of its growth management program. Furthermore, the State of Maryland itself recently enacted legislation to encourage "smart growth" land planning practices. To support this effort at the local jurisdictional level, the state has developed and made available a GIS database that incorporates parcel data, along with guidelines for a suggested land supply analysis methodology.

A third case study describes a recent effort to evaluate the supply of and demand for industrial land in the central Puget Sound region. Conducted by the Puget Sound Regional Council (PSRC) and the University of Washington Center for Community Development and Real Estate (CCDRE), this work seeks to improve methods of forecasting the future demand for industrial land. A small piece of a much broader set of regional growth management monitoring efforts, this case study offers methodological insights into the analysis of employment-based land uses, as well as the development of regional land management systems. In contrast to the other two cases, land monitoring in the central Puget Sound region is largely decentralized, causing this regional study to rely on land supply inputs from many local jurisdictions.

The three case studies are complemented by two appendixes. Appendix A is a detailed analysis of the findings of the survey of LSCM in jurisdictions nationally. Appendix B consists of summaries of land monitoring approaches in eleven selected jurisdictions. These cases further illustrate the various aspects of GIS and land monitoring discussed in the main chapters of this book.

BOOK STRUCTURE

The book has three parts. In Part I, the first chapter summarizes the state of practice and the challenges faced by land monitoring programs in this country. A second chapter introduces essential definitions and elements of land monitoring. It puts forward a generalized methodological framework for land supply monitoring and land capacity analysis, outlining the steps to be followed, variables to be considered, and critical issues. This chapter helps the reader conceptualize all of the elements of a land monitoring process.

Part II contains three chapters relating to the three principal case studies: Portland Metro, Montgomery County and the State of Maryland, and the Puget Sound Regional Council and University of Washington study. Chapter 3 reviews conceptual issues in land monitoring, business inventory methods as applied to monitoring, database design alternatives, and preferred units of analysis. It provides a framework for assessing Portland Metro's database and land monitoring approach. Chapter 4 discusses growth management and land supply monitoring programs in Montgomery County, as well as the State of Maryland Smart Growth Program and the MdProperty View GIS database. Chapter 5 explains the planning context for the central Puget Sound industrial lands study and discusses the methodological basis used to establish supply and demand for industrial land in the region.

Part III turns to thematic issues about the future of LSCM. Chapter 6 discusses technical advances and methodological responses to the need for accurate and timely land monitoring. Using the Portland Metro case for illustrative purposes, it focuses on conceptual issues in land use classification and tracking land use change over time. Chapter 7 describes inter- and intraorganizational conditions for cooperative efforts to use PBGIS for land monitoring. Chapter 8 reviews the potential role of urban land simulation models in refining the demand assumptions that inform estimates of land supply and development capacity. The focus is on UrbanSim, a disaggregated land use change model incorporating parcel-level data.

The chapters in Parts II and III include commentaries by experts in the fields of land use planning and GIS. Reflecting the nature of land use monitoring with GIS, these commentaries enrich the chapters' contributions by introducing different and sometimes even conflicting interpretations of the issues involved.

BASIC ISSUES

The use of parcel-based GIS in land supply and capacity monitoring generates many issues encompassing methodological approaches, the organizational and institutional parameters within which land monitoring takes place, and the promise and limitations of the technology itself. Two broad themes frame a discussion of the basic issues. First, land supply monitoring is a politically grounded activity that embraces the full array of interests engaged in the implementation of land use controls in general, and in growth management in particular. As such, the assumptions and criteria that define land "supply" and "capacity" are subject to continual debate and reassessment in the private and public arenas. Second, land monitoring is a data-driven activity that depends on an appropriate data collection, management, and analysis technology—for example, GIS, remote sensing, and spreadsheet and database management software. There is an evident dynamic in land monitoring between methodology and information system resources. Within these themes, five issues stand out. As they recur throughout this book, references are provided to pertinent chapters.

Issue 1: Defining Land Supply and Capacity

Land supply and capacity monitoring is an exercise in measuring the way land is used, and not used, in order to address the potential future use of land based on its physical and social attributes. It consists of much more than a running inventory of land utilization and commonly includes information on legal definitions of how land can be used, depending on environmental and physical factors affected by development, infrastructure levels, and land and real estate markets (Godschalk et al. 1986).

Decisions shaping the development of urban land involve a large number of participants and interests in a highly politicized process. Consequently, there are many definitions of the supply of land and its capacity (Dearborn and Gyri 1993; Bollens 1998). Naturally, the contentiousness of this process increases according to the relative scarcity of land deemed "developable." The need to establish "definition(s) of available land supply agreed on by both private interest and the government" (Godschalk et al. 1986) continues to be a major issue within land monitoring.

What constitutes demand for land is also a major area of contention. Although demand technically does not have to be considered in order to determine land supply and estimate its capacity, in practice, the adequacy of a given land supply is assessed relative to expected demand. Typically, this entails comparisons of aggregate supply and demand, such as dwelling unit capacity versus household increases, or square feet of nonresidential capacity versus employment growth. Other considerations of demand for land include consumer preferences for housing type, sectoral employment growth, and changes in construction and land development practices. As may be expected, there are significant differences between the public and private sectors about the time horizon over which the balance between supply and demand should be assessed.

The political nature of LSCM makes it subject to frequent, and indeed often sudden, changes. As a result, definitions of supply and capacity may change over time, reflecting both methodological shifts and the ongoing influence of the different interests involved in the process.

In Chapter 2, a framework for LSCM is proposed that builds on current thinking and best practices in the field and provides a consistent conceptual structure and a related set of definitions.

Issue 2: Measuring Land Supply and Capacity—Negotiated Processes Based on Plans and Data

In principle, approaches to LSCM do not have to rely on either large data sets or sophisticated models of urban development. Indeed, many cases of current practices surveyed do not. Instead, they rest on gross, aggregate inventories of land available for development—primarily vacant land at the fringe— projecting its full build-out capacity under current zoning and commonly adjusting these totals via areawide percentage deductions for land deemed "unavailable" for development. However, such efforts provide only rough estimates of land supply, generally on a one-time basis, and do not constitute an adequate foundation for ongoing "monitoring" of land supply and development.

For land monitoring purposes, establishing the appropriate criteria to define land supply and capacity *and* devising ways to quantify land capacity according to these criteria are delicate and difficult tasks. For example, categories such as "developable" or "redevelopable" land may seem, at the outset, to be simple to establish. Yet, for instance, although defining vacant land as that without improvements is easily achieved, it is more problematic to establish what constitutes vacant land that is "developable." Questions pertaining to the determination of sufficient levels of urban infrastructure needed to make land developable illustrate the potential complexities involved in making this determination: Is land developable when it is not serviced? What level of service is necessary for vacant land to be developable? How is the geographic extent of serviced land actually measured? What services are considered essential for developability—sewer, water, parks, schools, transportation infrastructure? By what criteria is unserviced land deemed "serviceable"? Is serviceable land, land that is "planned" to be serviced? According to which plans? Or is serviceable land that which private sector developers are willing to service at their own costs?

Negotiating the complexities of defining land supply and capacity is a process that increasingly involves input, and usually some measure of agreement, from many stakeholders. To be successful, this negotiated process depends on two elements: (1) a framework provided by a set of *plans* defining how land should be used in the future and (2) an *information base* about the land itself, including all of the factors that can influence its use. The most sophisticated LSCM systems today have evolved through the give-and-take between the need for better information on which to base and evaluate land use plans, and the development of increasingly precise data on land supply and use. Thus, the collection and analysis of land data are increasingly central to the plan-making process. This presents a new situation for planners. In order to meet heightened requirements for accurate and detailed land supply information, planners are incorporating computerized analysis tools, particularly GIS, and drawing on land-related data that have traditionally belonged to other domains of local governance.

The measurement of land supply and capacity is addressed in all of this book's chapters, but particularly in Chapters 2 through 6.

Issue 3: Land Information Systems

An LSCM is highly data dependent. It is also influenced by the needs of numerous and varied stakeholders in the database development project. Common data inputs to the process include current and planned land uses and intensities, zoning and other regulations, environmental constraints (usually as defined by regulation), transportation and utility infrastructure, urban facilities and services (parks, schools, etc.), and land ownership and valuation. Newly emerging data requirements include land price and indicators of market availability, intended both to improve the accuracy of analysis results and to satisfy private sector requests that supply estimates accurately reflect market realities. The emergence of increasingly sophisticated land information systems—designed to serve a range of local government functions—presents opportunities to leverage their use for monitoring land supply. Yet because these systems are established within, and traditionally utilized for, local land management and taxation departments, capitalizing on them entails building bridges between disparate professional cultures and institutional mandates. However difficult, such linkages will be essential for the future evolution of LSCM.

Chapter 7 explores these issues. It outlines the actors and functions that must be included in land monitoring database development and focuses on organizational and institutional settings and the potential and constraints they present in establishing the necessary linkages.

Issue 4: Spatiality of LSCM

Measures obtained in land monitoring are highly dependent on the *location* and geographic *scale* at which supply and capacity are sought. The task is not only to define how much land is in the supply, but also where it is. Spatiality is of increasing importance in land monitoring because the spatial dimensions of land and its attributes can now be captured over large areas and in detail.

The first spatial concept to address is the *geographic extent of the analysis*—the area over which data are collected and supply and capacity are analyzed. This area may be an urban region, a city, a planning or analysis subarea, or a neighborhood. Geographic extent ideally relates closely to demand forecasts, which are usually regional. Monitoring land supply for smaller areas within larger regions is common, but the value of area-based monitoring is limited if undertaken in isolation, without regard for the interactions of supply and demand in other areas. The size of the study area also implies an appropriate or feasible scale or level of disaggregation; regional analyses are generally carried out at a coarser resolution than analyses of local jurisdictions or planning subareas.

A second spatial concept concerns the *level of detail* of the data used to establish supply, and the level of resolution at which capacity is analyzed (whether, for example, it is done at the level of the parcel, the analysis zone, or the particular jurisdiction). The spatial units of data collection and spatial units of analysis determine the level of precision and locational specificity of the results.

The choice of spatial unit involves access to new types of data and incorporation of new criteria for the analysis. Historically, spatial data on land uses have been available in relatively large units (representing contiguous land use areas or involving coarse grid cell classifications), which do not relate well to the scale of data describing land development events. However, as parcel-based GIS become more common, data on land ownership, rights, valuation, parcel size and shape, and adjacencies can be more easily captured and analyzed than ever before. Indeed, the availability of PBGIS to local and regional planning agencies has greatly increased the level of disaggregation at which land monitoring can be done. Specifically, by allowing land monitoring to take place at the scale of the tax lot, PBGIS enable crucial links to be established between planning functions, such as long-range land use planning and land use regulation, and market-based land development—two arenas that have notoriously been separated in practice. Because parcel data match the scale of the development process, PBGIS, in principle, enable analyses of supply and capacity to take into account short-term market variables and development practices (Dueker and DeLacey 1990), as well as the potential for infill and redevelopment.

As always, however, the coin has another side, and several caveats are in order. First, greater precision of measurement and disaggregation of analysis do not necessarily lead to greater accuracy, because the quality of the results depends on the quality of the data themselves. Assessments of the use potential for each individual parcel may not be valid if current land uses are not recorded accurately at this level. Second, parcel-level analyses—that, for example, identify parcels that are "developable"—cannot be considered as *predictors* of the future development potential of each and every parcel, because the future decisions of owners and developers remain uncertain. It follows that the criteria used for developability do not (and cannot) include the specifics that make each parcel unique. Finally, as many determinants of land use may not fit a particular parcel (for example, environmentally sensitive land), multiple units of data analysis are preferable.

As a basic rule, a distinction should be made between the spatial units of data *collection* and the spatial units of *analysis*. Data collection usually relies on multiple spatial units (including parcels, environmental features, census tracts, and infrastructure service areas). Units of analysis, on the other hand, largely depend on planning and policy requirements, the capabilities of the analytical tools available, and the specific type and purpose of the analysis being performed (such as a vacant lands inventory or a redevelopment study). Units range from parcels to analysis zones, to zoning districts or planned land use classifications.

Issues of spatiality are reviewed throughout this book, but specifically in Chapters 1 through 4, 6, and 8.

Issue 5: Temporality of LSCM

The term "monitoring" clearly invokes the importance of temporality in LSCM. Temporality refers to a set of simple, yet essential, questions encompassing (1) *when* land supply and capacity are established, (2) *how often* they are established, and (3) *within what time frames*. These questions, in turn, relate to the characteristics of the data—their timeliness, the capture of longitudinal data, and the form in which these data are available.

As a first consideration, land monitoring practice has generally been linked to long-range planning, assessing supply availability and land capacity over *long time periods* (typically 20 years) corresponding to the traditional time horizons of comprehensive plans. Compounding the challenges associated with addressing long time periods is the static nature of many of the assumptions guiding analyses of supply and capacity. For example, few, if any, significant changes in land development practices are generally anticipated, even over the long run. In contrast, actual land use change often takes place relatively rapidly—typically within building booms, which last only a few years. Real estate cycles are notoriously difficult to predict and, hence, difficult to take into account in public policy. For the private sector, however, these shorter-term trends in supply and demand are essential, and often all that really matters. Thus, the development and real estate industries typically operate within time spans of a few months to a few years.

Clearly, LSCM is highly dependent on the time frame to which it applies. Determining the amount and location of land available for development within 5 years is quite different from establishing the same information for a 20-year period. Short- to medium-term supply and capacity analyses are useful in two specific contexts. One area, as noted, is the functioning of land markets, which encompasses assessments of both market availability and development feasibility. Although these factors are difficult to assess, they will have significantly restrictive impacts on short- to medium-term supplies (Godschalk et al. 1986).

The second area in which these analyses are useful is infrastructure planning, which typically addresses the medium term—capital improvement plans (CIPs) are generally updated every 3- to 7-years. Without an established framework for consistency between these plans and 20-year comprehensive planning efforts, long-term land monitoring has been of only moderate importance to this function. At the same time, however, the introduction of federal and state requirements for linking land use and infrastructure planning (as, for instance, in the Intermodal Surface Transportation Efficiency Act (ISTEA) and the Transportation Equity Act for the 21st Century (TEA-21) legislation and in various state-level consistency requirements for local comprehensive planning) renders the information generated by land supply monitoring increasingly relevant and useful for planning and managing capital facilities and budgets.

As a second consideration, LSCM activities have been notably sporadic and discontinuous. The "snapshot" approach is typical in land supply monitoring. It reflects the tradition of making land use and supply inventories in periodic cycles corresponding to comprehensive plan updates. These updates are often carried out at irregular intervals, relying on episodic data collection (from aerial photography, windshield surveys, etc.). The snapshot approach clearly fails to fulfill the promises of an ongoing land supply "monitoring" and does not provide adequate and timely information either to public agencies involved in infrastructure planning or to the private land development sector.

Parcel-based GIS again open up interesting opportunities in this area, particularly when implemented within distributed networks linking multiple participants. First, the parcel layer provides the critical link between traditional land records and specific development and land use change "events" (such as permit applications, occupancy certificates, and land transfers). Second, these events can likely be captured through "transactional" updating linked to the administrative land management "actions," such as permit approvals or rezones. And third, with the responsibility for data updating shared among multiple agencies and departments, "real-time" land supply monitoring can theoretically be achieved. Indeed, new technologies have spurred interest in the possibility of building systems to inventory land on an *ongoing or continuous basis*. A proposal for such a system is presented in Chapter 3 in relation to the Portland Metro case study. Data models and technical requirements for such an approach are reviewed in Chapter 6, along with a broader treatment of issues related to the management of longitudinal GIS data in land information systems. Local governments, not least among them the planning departments themselves, have traditionally not been oriented toward capturing or working with historical data. Even with the introduction of frameworks and incentives for change in this area, the barriers erected by professional habits and institutional priorities will likely continue for some time. Chapter 7 discusses how organizational structures can strongly influence the practice of land monitoring at a technical level.

THE FUTURE

The aforementioned issues, as well as many other less conspicuous issues, are addressed in this book. Generally, the work that follows shows that land monitoring is becoming an increasingly sophisticated and central activity within the realm of urban and regional planning. As such, new land monitoring approaches and techniques enable planners to approximate the dynamic reality of land use change. They greatly help to address, directly and effectively, concerns about the multiple impacts of urban growth. Technological advances are also working to narrow the gap between definitions of land supply and capacity coming from the public and the private sectors. Anticipated technological breakthroughs may bring the views of the two sectors closer in the future. Yet perhaps the biggest question about the future is the ability of public-sector agencies to develop organizational structures that allow them to sustain the level of support (technical and financial) necessary for maintaining and improving the excellent land information systems and LSCM programs now being developed.

Finally, because parcel-based land monitoring with GIS is only emerging, this book emphasizes general methodological and technical considerations of tracking and analyzing land supply and capacity. It provides planners and policymakers with an assessment of where LSCM is and where it needs to go. To accomplish this aim, the book casts a broad net, encompassing a full range of growth management tools as well as more general land use planning concerns as they relate to LSCM. However, being "technical" in nature, this book largely stops short of addressing the state of land supply itself in this country, or the contentious, even potentially explosive, urban land policy questions on which the "meaning" of that supply hinges. It is crucial to remember that LSCM is but a means to many ends. GIS offers the potential to project a clear and precise picture of the interaction between development regulations, on one hand, and such timely policy issues as transportation investments, central city revitalization, and environmental protection, on the other. A fully realized use of GIS technology will likely help to place land supply issues at the forefront of policy. In so doing, it will add fuel to the growing national debate on the sustainability of current urbanization patterns and the public and private practices that create them. This, no doubt, will be the subject of future publications on land supply and capacity monitoring.

REFERENCES

American Planning Association. 1998. *Growing Smart legislative guidebook*. Phase II. Interim ed. Chicago: American Planning Association.

Bollens, Scott A. 1998. Land supply monitoring systems. In *The Growing Smart working papers*. Vol. 2. Chicago: American Planning Association.

Bollens, Scott A., and David R. Godschalk. 1987. Tracking land supply for growth management. *Journal of the American Planning Association* 3: 315–327.

Chrisman, Nicholas. 1997. *Exploring geographic information systems*. New York: John Wiley & Sons.

Crane, Ed. 1997. Telephone interview by author, October 24.

Dale, Peter F., and John D. McLaughlin. 1988. *Land information management: An introduction with special reference to cadastral problems in third world countries.* Oxford: Clarendon Press.

Dearborn, Keith W., and Ann M. Gyri. 1993. Planner's panacea or Pandora's box: A realistic assessment of the role of urban growth areas in achieving growth management goals. *University of Puget Sound Law Review* 3: 975–1024.

De Grove, John M. 1992. *Planning and growth management in the states.* Cambridge, Mass.: Lincoln Institute of Land Policy.

Downs, Anthony. 1994. *New visions for metropolitan America.* Washington, D.C.: Brookings Institute; Cambridge, Mass.: Lincoln Institute of Land Policy.

Dueker, Kenneth J., and P. Barton DeLacey. 1990. GIS in the land development planning process: Balancing the needs of land use planners and real estate developers. *Journal of the American Planning Association* 4: 483–491.

Dueker, K. J., and Daniel Kjerne. 1989. *Multipurpose cadastre: Terms and definitions.* Falls Church, Va.: American Society for Photogrametry and Remote Sensing and American Congress on Surveying and Mapping.

Enger, Susan C. 1992. *Issues in designating urban growth areas.* Parts I and II. Olympia, Wash.: State of Washington, Department of Community Development, Growth Management Division.

Godschalk, David R., Scott A. Bollens, John S. Hekman, and Mike E. Miles. 1986. *Land supply monitoring: A guide for improving public and private urban development decisions.* Boston: Oelgeschlager, Gunn & Hain, in association with the Lincoln Institute of Land Policy.

Huxhold, William. 1997. Telephone interview by author, October 11.

Jacobs, H. M. 1988. *Land information systems and land use planning: An annotated bibliography of social, political and institutional issues.* CPL Bibliography Series, no. 208. Chicago: Council of Planning Librarians.

Kaiser, Edward J., David R. Godschalk, and F. Stuart Chapin. 1995. *Urban land use planning.* Urbana and Chicago: University of Illinois Press.

Kivell, Philip. 1993. *Land and the city: Patterns and processes of urban change.* London and New York: Routledge.

Klosterman, Richard. 1997. Planning support systems: A new perspective on computer-aided planning. *Journal of Planning Education and Research* 17: 45–54.

Kollin, Cheryl, Lisa Warnecke, Winifred Lyday, and Jeff Beattle. 1998. Growth surge: Nationwide survey reveals GIS soaring in local governments. *GeoInfo Systems* 2: 25–30.

Peirce, Neil. 1997. Maryland's "Smart Growth" law: A national model? *Washington Post Writers Group page.* <http://www.clearlake.ibm.com/Alliance/newstuff/peirce/peirce_042097.html> (October 21, 1997).

Part I

Overview

1

Current Land Monitoring Practices and Use of GIS: Challenges and Opportunities

Anne Vernez Moudon and Michael Hubner

The monitoring of urban land has taken place in a variety of contexts in this country. As a general activity, it seeks to record changes in land use, to track land development, and to analyze the patterns and conditions that emerge over time. Specifically, land supply and capacity monitoring (LSCM) focuses on the supply of buildable land and on the capacity of that land to accommodate future development. As such, it also serves to assess future potential uses of land, especially in relation to how zoning and other regulations support or constrain urban expansion and concentration.

The land monitoring process, in its most developed form, is a dynamic and comprehensive activity carried out on a continuous basis. Holmberg (1994, 8, with reference to Masser [1984]) characterizes monitoring as "crucial for any successful handling of the territorial concern . . . [exhibiting] both similarities and differences compared with traditional census, surveying, and mapping, [and comprising] continuous and real-time supervision of a set of variables within a geographic region." Furthermore, monitoring extends beyond passive

Monitoring Land Supply with Geographic Information Systems, edited by Anne Vernez Moudon and Michael Hubner ISBN 0 471371673 © 2000 John Wiley & Sons, Inc.

description to include active intervention in response to information generated. These observations describe well a fully implemented system for LSCM.

Urban and regional planners bear the chief responsibility for land monitoring. Although not a widespread activity, LSCM has emerged as a distinct area of planning in metropolitan regions where the demand for land is high, thus supporting the political will to restrict the effective supply of land available for urbanization through policies and regulations that limit development, particularly at the urban fringe. As such, LSCM stems primarily from efforts to manage urban growth.

Concerns about the supply of urban land have become increasingly urgent over the past decade. During this period urban growth has been characterized by rates of land consumption substantially higher than increases in population. Interest in the supply of urban land springs from two separate yet related considerations. One is the conservation of open space—agricultural and natural areas—within metropolitan regions, particularly at the urban fringe. The other relates to the high costs of providing infrastructure (especially transportation infrastructure) required to serve sprawling metropolitan development. As these issues have come to the center stage of local political agendas, many jurisdictions have passed laws to manage growth and to monitor closely changes in urban land use and available supply (De Grove 1992; Stein 1993). Generally, growth management mandates aim for orderly and efficient growth patterns, not only to preserve land outside the urbanized area and to reduce infrastructure costs, but also to mitigate the negative impacts of growth on traffic flow and air and water quality and to distribute the differential changes in local tax bases that result from uneven development (Diamond and Noonan 1996). As an accompaniment to growth containment policies, LSCM serves to assess the adequacy of the land supply (limited as it is by regulation) to accommodate future population and employment increases, while achieving the objectives of these policies.

Although LSCM has become an important component of urban planning within the context of growth management, it has yet to emerge as a bona fide specialization within planning study. Research on the subject is limited, and no comprehensive theory has been articulated to guide its practice. Few academic researchers in planning claim land supply monitoring to be their area of scholarly activity. Godschalk's and Bollens' work in the mid-1980s (Godschalk et al. 1986; Bollens and Godschalk 1987) still stands alone in its comprehensive coverage of issues and methods related to LSCM. As Bollens noted in summarizing the field for the American Planning Association (Bollens 1998), various researchers have addressed separate issues that relate to specific aspects of land supply and capacity—such as urban sprawl, housing and land prices, costs of infrastructure, regulatory impacts, and program evaluation. Little work has focused directly on the technical, methodological, organizational, or substantive aspects of LSCM as a distinct planning activity. This gap may be partly explained by a lack of "critical mass" of well-developed LSCM programs. Another explanation may lie in the intrinsic complexity of LSCM, which must address an increasingly wide range of interacting variables, including environmental conditions, capital facilities and infrastructure planning, regional economies and land markets, and real estate and development practices.

At this point, therefore, land supply monitoring in this country is defined more by a set of practices than by a cohesive theory or body of empirical evidence. A handful of jurisdictions pioneered LSCM in the 1970s and 1980s, among them the Metropolitan Council (Minneapolis-St. Paul), Minnesota; Montgomery County, Maryland; Portland Metro, Oregon; Lane Council of Governments (Eugene-Springfield), Oregon; and the San Diego Association of Governments, California. (The work of these jurisdictions is updated in Appendix B.) These and other pioneers have essentially determined the state of the practice. Although jurisdictions that conduct LSCM attempt to keep abreast of what others are doing relative to land monitoring, there are no formal ties between them—for example, there is no national group or organization that keeps tabs on ongoing developments in the field. As a result, the various approaches to LSCM still await critical review and systematic evaluation and comparison.

This chapter summarizes the characteristics of current practice in land supply and capacity monitoring in the United States. The summary relies on research of more than four dozen jurisdictions across the country, a subset of which are analyzed in detail in Appendix A. In this chapter the general scope of practice is reviewed first, followed by a discussion of the recent advances in GIS and their implications for LSCM. (The actual methodological dimensions of LSCM are addressed in Chapter 2.) The chapter concludes with a summary of the challenges and opportunities facing the field now and through the coming decade.

LSCM PRACTICES

The term *land supply monitoring* suggests an activity whereby the supply of land is tracked as it is occupied by urban activities, and as changes take place in the use of urbanized and urbanizing land over time. It also suggests a systematic assessment of potential uses of the land supply for the purposes of informing land policy. In the reality of practice, LSCM is rarely implemented in a comprehensive way. Few, if any, jurisdictions have managed to establish monitoring systems as ongoing inventories of the entire land supply. Nor have they used LSCM as a tool to measure on a regular basis the effectiveness of established policies and regulations and the possible impacts of new regulations, policies, or land development practices.

A survey of jurisdictions across the country shows that few have access to databases that are comprehensive or detailed enough to address the entire range of urban land uses. In addition, few systematically monitor trends by capturing longitudinal data on land use and development. Hence, most monitoring approaches involve partial inventories carried out as "snapshots" of the land supply, as opposed to continuous processes of data collection and analysis. Furthermore, these snapshots take place periodically and not always at regular intervals.

The limited adoption of LSCM among jurisdictions nationally has many explanations, of which several stand out. First, LSCM is often prohibitively expensive to carry out, especially in terms of database development and maintenance. Second, most jurisdictions have not implemented growth management policies, which has resulted in uneven institutional support and requirements for LSCM. And third, in jurisdictions where growth management *is* in place, the contentiousness of local politics may create an unstable political climate for sustained LSCM system development.

Characteristics of Practice

Our survey of jurisdictions (see Appendixes A and B) highlights six general characteristics of current LSCM practice. First, it is led by the public sector and typically carried out by local governments—municipal, county, and metropolitan jurisdictions. At the same time, land monitoring activities are usually closely watched by the private sector. Second, it is regional in scope, reflecting the metropolitan-wide reach of land markets, major infrastructure systems, and growth control policies. Third, it addresses primarily medium- to long-term supply and capacity prospects, largely to relate appropriately to comprehensive plans and to regional growth forecasts, which usually extend over periods of 10 to 20 years. Being closely linked to long-range planning, LSCM is most successfully performed as a centralized or coordinated activity led by a metropolitan planning organization (MPO) or similar metropolitan-wide body. In this context, LSCM has been successfully employed in evaluating the implementation of local and regional plans. It has contributed to keeping issues of urban growth, land consumption, and infrastructure development current, and to raising peripherally the public awareness of the environmental and fiscal costs of sprawl.

Fourth, LSCM involves the collection, maintenance, and analysis of large, complex data sets, with elements ranging from aerial imagery to infrastructure networks to parcel-specific land records. As a multifaceted and highly data-dependent activity, LSCM generally requires significant coordination to complement extensive investments in specialized labor and technology by local government. "Corporate" data centers are common, functioning as the equivalent of line departments responsible for generating, maintaining, and processing data for other departments within a contributing jurisdiction or jurisdictions. Data are captured and stored for various units of land, corresponding to data type and administrative function (e.g., parcels for assessments, districts for utility servicing). Analyses of the data range widely in the level of aggregation with which they address supply and capacity. Although parcel-level analyses are increasingly common, many efforts aggregate data to a zonal level—

such as census tracts, Transportation Analysis Zones (TAZs), Forecast Analysis Zones (FAZs), or zoning districts—in order to evaluate the potential for land to be developed. They yield results that are fairly aggregated and, as discussed later, largely inadequate to meet the current needs of growth management policy and urban land and systems management.

Fifth, most land monitoring programs have been sporadic rather than continuous efforts, whereby land supply and capacity are analyzed for an established time horizon but often not tracked over time to support a good understanding of trends. Although a comprehensive, one-time inventory can provide a baseline against which to measure trends and subsequent changes, capturing longitudinal data over any significantly long period has proven difficult. Moreover, changes in technology, inconsistent financial support for data collection and maintenance, and shifting mandates for long-range planning have all contributed to the lack of continuity in monitoring efforts nationally.

Sixth and last, our research reveals the following primary substantive orientations within LSCM practice:

- *A concentration on the development potential of land reserved for single-family residential use.* The focus of LSCM on single-family land supply, at the expense of other residential or nonresidential land uses, is typically justified because such use is the largest consumer of land for urbanization.
- *A focus on vacant land.* LSCM centers on the consumption of raw land at or beyond the urban fringe because it is seen as the leading threat to open space, resource lands, and natural areas, as well as to the cost-efficient provision of new infrastructure and services. As such, the supply of vacant fringe lands is a major factor in ongoing debates over urban growth containment policies.
- *Limited treatment of employment-based uses.* A handful LSCM efforts have recently begun to focus on land supply for industrial, commercial, and office uses. Initial evaluations of existing capacities for such uses have generally found them to be ample. However, accurately estimating future development potential for employment-based land uses is problematic. The relationship between job growth and land needs has been found to be far less linear than that between population growth and land needs. In addition, findings of employment capacity studies are frequently criticized by local development and business interests, especially when the location of the "available" lands does not correspond to historical patterns of demand for these uses.
- *Limited treatment of multiple or mixed uses.* These types of development have also only recently been addressed as a factor in land capacity. This too poses challenges to reliable assessment, owing to the dynamics of market preferences, development practices, and regulations allowing or encouraging mixed-use projects.
- *Limited treatment of lands available for infill or redevelopment.* Relatively few jurisdictions have assessed in detail the potential of land already

urbanized. Such efforts have been impeded by a lack of appropriate data (related, for example, to site-level land use and improvements), as well as by methodological constraints (especially in establishing acceptable and reliable criteria for identifying lands with redevelopment and infill potential).

- *Piecemeal consideration given to development constraints related to environmental factors (wetlands, floodplains, steep slopes, etc.).* When environmental constraints are considered in LSCM, they are typically defined to reflect prevailing limits imposed by regulations. Less commonly addressed are partial constraints on site development or comprehensive carrying capacity limits.

- *Definition of serviced and serviceable land limited primarily to transportation infrastructure.* In most LSCM approaches, sewer, water, and other utilities come second in order of importance, and schools, parks, and recreational facilities remain a distant third consideration.

- *Reliance on status quo assumptions about future development.* Monitoring programs have not addressed well the potential impacts of new or innovative land development approaches, new infrastructure or building designs, or new construction practices. Local jurisdictions have thus yet to use land capacity analysis techniques to explore aggressively the potential of the land supply to be developed in more efficient, resource-conserving, or cost-effective ways.

Private-Sector Concerns

LSCM has, as may be expected, been subject to the contentious climate of growth management politics. In debates over "planning for" versus "planning with" growth, critics of the former approach generally assert that regulations limiting development or containing urban growth excessively restrict both the amount and location of buildable lands. Home builders and other development interests also question the timing of land availability, particularly short-term supplies of buildable single-family lots and large sites for multifamily development and nonresidential uses.

Land monitoring programs focused on long-range supplies of land over entire urban regions tend to overlook factors that affect short-term land availability—even though they may be the target of private-sector criticism. Over the short term, regional inventories typically fail to capture locational variations in market demand and may have difficulties addressing institutional factors such as permit approval rates and staged extensions of infrastructure. Moreover, their treatment of physical and environmental constraints often is too general or too spatially coarse to address specific barriers to development. Finally, regional inventories typically lack detailed consideration of various economic variables, such as the demand for large parcels (particularly for industrial uses), market availability, land prices, and the location preferences of households and firms. Negotiated within the political arena of growth management, these criticisms have had an impact on LSCM, spurring efforts to assess land supply in increasing detail and incorporating an increasingly complex array of constraints and market considerations.

Recent advances in land information technology, and specifically the implementation of parcel-based geographic information systems (PBGIS) by local governments, hold some promise that future land monitoring will be able to respond more directly and comprehensively to many of these concerns in addition to meeting the land policy analysis needs of public decision makers. The following sections review the state of the art in geographic information systems (GIS) and discuss the implications of the new technology for LSCM.

ADVANCES IN LAND INFORMATION TECHNOLOGY

Land record keeping has been the subject of elaborate bookkeeping systems since the beginning of urbanization. Historically, advances in writing and printing technology were quickly applied to improve the state of land records. Computers have naturally followed suit as a welcome aid to maintaining land information. Even in computerized form, however, land information was, until recently, stored and retrieved in two largely separate formats: land records in tabular form and maps depicting the land parcel (or other areas of land) showing its location and physical extent. The latest trends in using GIS for managing land information, and especially parcel-based GIS are having a significant impact on land monitoring. Following are a review of recent developments in land information technology used by local government and a discussion of the potential of PBGIS to further enhance LSCM.

GIS in Local Government: A Decade of Expansion

The use of computerized land information systems (LIS) for land supply monitoring from the mid-1970s through the mid-1980s was reviewed in Godschalk et al. (1986). Drawing on their survey of the practice, the authors described a prototype automated land supply information system (ALSIS), consisting of a set of computerized databases for tracking both parcel-based land records (a "parcel file") and in-process development and development applications (a "project file"). Using case studies, the authors illustrated the implementation of various specific components of such a system in a range of local and regional government settings.

At the time of this study, geographic information systems (GIS) technology was in its infancy as a tool for local land planning and management; only 6 of 24 jurisdictions surveyed by Godschalk et al. (1986) had developed computer-mapping capabilities. Since these observations were made, more than ten years ago, GIS has advanced considerably and its use by local and other governments has mushroomed, extending from daily routine tasks to sophisticated special-purpose analyses. As part of this trend, planning departments have become primary users of GIS. The rate of GIS usage is especially high among jurisdictions that practice some form of land supply monitoring. Of the more than three dozen jurisdictions surveyed for this work that actively practiced LSCM, nearly all utilized GIS as a tool for capturing, managing, or analyzing land data, and more than half did so with a parcel layer as a major component (see Appendix A).

Various technological changes have fostered the adoption of GIS by local governments, enhanced its application for multiple purposes, and increased the affordability of system implementation. First, distributed computing environments, incorporating local area networks and network file systems, have increased flexibility and reduced redundancy related to both data storage and software. Second, the introduction of personal computers (PCs) and personal workstations powerful enough to process large databases, have enabled the storage and manipulation of highly detailed and extensive spatial data layers, such as parcels. Third, relatively low-cost, user-friendly, and increasingly powerful desktop software has allowed a broad range of local government staff to utilize GIS on their own computers (Public Technology Inc. et al. 1991; Korte 1997; Orman 1997).

A wide range of benefits accrue from GIS adoption and make it an attractive investment for local governments. These benefits include (1) increased effectiveness of urban services delivery (e.g., utilities management, emergency services), (2) cost savings through the automation or streamlining of routine tasks (especially custom mapping and on-demand data retrieval), (3) increased employee productivity, (4) reduction of redundancy between and within departments (especially related to the creation and updating of spatially referenced data), (5) revenues gained from selling digital GIS and map products, (6) improvements in the speed and accuracy of responses to queries, and (7) multiple benefits stemming from improved decision making at all levels (Public Technology Inc. et al. 1991; Korte 1997; Klosterman 1997).

A recent national survey of cities (population greater than 25,000) and counties (population greater than 50,000) bears out the dimensions of the spread of GIS to local jurisdictions and to a wide range of departments and functions within them (Kollin et al. 1998). The survey results show that in 1996, 77 percent of respondents (85 percent for cities of more than 100,000 population) reported using GIS. Among GIS users, the average number of departments using GIS was 2.34 (2.85 for large cities). The most common uses of GIS included comprehensive planning, zoning and subdivision review, transportation planning, and utilities and storm water management. The types of GIS data most often utilized included road networks, political and administrative boundaries, hydrologic features, land use and zoning maps, and cadastral/land records (all more than 80 percent usage in 1997). Interestingly, surveyed jurisdictions with or without GIS most often identified the impacts of "growth and development" as a top "natural resources" concern. However, among these respondents, only 42 percent indicated that they actually used

GIS to address growth and development problems. This last finding suggests that GIS, in its current stage of implementation within local government, presents a largely untapped resource to address problems of urban growth, in part through the monitoring of land supply and capacity.

As GIS technology has spread, so has the creation of and access to spatial and spatially referenced data. One of the most significant developments for land monitoring has been the addition of parcel layers by many county and municipal GIS programs. In a separate study of 29 local jurisdictions with operational GIS in the southeastern part of the United States, Nedovic-Budic (1993) found that parcels constituted the most frequently mapped feature. Often spearheaded by an assessment or engineering department, the construction of a digital parcel base map can incorporate older computerized land information systems (LIS) data (which are based primarily on the parcel record) and provide new tools for the display and analysis of land-related data in a spatially explicit manner. Such land records "modernization" is now in progress widely throughout the nation, and parcels are quickly becoming standard elements in the GIS of most sizable local jurisdictions (Huxhold 1997; Kollin et al. 1998). The following sections address specific implications of this trend for the application of GIS to the monitoring and analysis of land use, supply, and capacity.

Parcel-based GIS: Representation and Implementation

To understand the potential of parcel-based GIS for land monitoring, it is useful to consider the several different ways that parcels may be represented spatially in a GIS—as points, polygons, or grid cells—and the appropriate application of each approach. The simplest representation is a parcel-point layer, composed of geocoded parcel locations that correspond to either parcel centroids (xy coordinates) or to "address ranges" along a street network. The parcel-point approach is widely used in the private sector for demographic and market analysis, for site searches, and for the marketing of development and real estate products (such as with computerized Multiple Listing Service data) (Castle 1993, 1997; Thrall 1997). Within the public sector, parcel-point GIS has been and continues to be used for a variety of planning analyses and land monitoring applications, especially by jurisdictions in which a fully digitized parcel coverage is not yet available. However, parcel points present significant limitations for accurate land monitoring, partly because of the lack of precision in geocoding, especially in address matching (Drummond 1997), and because of the absence of a spatially extensive representation of the boundaries and area of the parcel. Parcel points thus limit the ability to perform accurate overlays and to analyze land supply within a detailed geographic context (i.e., the parcel itself and its surroundings).

For the multiple purposes of land information management in local government, parcel-point coverages are often utilized as a stopgap solution before the completion of a true parcel-based GIS (see Dueker and DeLacey (1990) on the County Geographic Index, and the case of Snohomish County in Appendix B). A true parcel-based GIS incorporates a fully digitized parcel layer composed of parcel boundaries, allowing parcels themselves to be represented as discrete areas (polygons). As such, the parcel layer offers many advantages over parcel points, including enhanced display and map production and the ability to consider precise site-level details necessary or desirable for a variety of analytical operations. The conversion of parcel maps to GIS format is generally accomplished either through digitization of lines on paper maps, or as constructed directly from deeds and survey information using coordinate geometry and aided by alignment with planimentric or orthophotography layers (Donahue 1994). The parcel layer may also be derived partly from previously developed digital computer-aided design (CAD) drawings (commonly used by public works and engineering departments since the early 1980s).

Whether employing a GIS representation of parcels as points or as polygons, the creation of a parcel layer entails the assignment of a unique parcel identifier, usually referred to as a parcel identification number (PIN), used to link the spatial features to individual land records and other parcel-specific data (Huxhold 1991). As a secondary, but potentially useful, parcel identifier, a standardized Master Address File can link parcels to site-level data produced by the numerous local government departments that identify parcels by address rather than PIN in their record-keeping systems (Donahue 1994).

Point or polygon representations of parcels are features of vector GIS. Parcel polygons may be converted to grid cell format or raster GIS. Recent advances have increased the ease of data conversion from vector to raster and vice versa. Although some detail may be lost through this process, grid cells of 20 to 50 feet are usually adequate to capture the underlying parcels, with at least one cell corresponding to each parcel polygon. The advantages of working with parcels in grid cell format include: (1) reducing the computational burden of large-area parcel-level calculations and transformations, especially those relying on multiple geographic layers, (2) improving the ability to consider weighted variables in screening for site suitability (Dueker and DeLacey 1990), and (3) allowing analysis of the potential use of parcels by considering their small-scale geographic context—such as with "neighborhood" calculations that assign values to cells based on the values of nearby cells. (See Portland Metro and Lane Council of Governments case summaries in Appendix B for examples of raster GIS analysis.)

Within urban and regional planning, most land monitoring applications of GIS without a parcel coverage (or parcel layer) have been, and continue to be, based on zonal geography.[1] Most frequently, zones correspond to adminis-

---

[1] Two other planning applications of GIS that relate indirectly to land supply monitoring deserve mention. First, the long-standing use of GIS for environmental planning and analysis relies primarily on raster representations of environmental variables, and is conceptually concerned more with surfaces, such as land cover, than discrete areas. Second, the use of GIS for transportation planning has emphasized network analysis and has only secondarily used zonal units, such as TAZs, to represent travel demand by general location.

trative or statistical areas—census tracts, block groups, and blocks; planning areas; and Transportation Analysis Zones (TAZs) and Forecast Analysis Zones (FAZs)—allowing planners to relate land areas to sociodemographic and economic statistics that are collected at or aggregated to specific zonal units. These practices derive from the early use of census data (when it was available only in tabular form) and the tradition of social area analysis. Zones may also relate explicitly to land use and development, including regulatory units (e.g., zoning districts, planned land use areas) and areas representing contiguous land uses (typically classified from aerial photography). Although parcel-level land records can be aggregated to any of the various zone types for subarea analysis of supply and capacity, the relatively coarse spatial resolution of zones, as well as their abstract delineation, imposes limitations on the utility of analysis results generated with zonal GIS.

The coarseness of zonal geography (TAZs and FAZs that encompass several thousand acres are common in suburban areas) limits the degree to which the patterning and clustering of development and development potential can be detected.[2] PBGIS, representing parcels that range from fractions of acres to multiple acres, even in suburban areas, offer a more precise basis, in terms of scale and level of disaggregation, than zonal GIS for identifying and analyzing spatial patterns. In addition, because the parcel corresponds directly to many of the administrative and private-market development decisions that accompany incremental land use change, it has been recognized as the preferred unit for land monitoring (Bollens and Godschalk 1987; Enger 1992; Vrana and Dueker 1996; Bollens 1998).

Typically, however, additional land supply data are captured and represented in units other than parcels. Over the past decade, theory and practice of GIS for land management and planning have expanded the concept of a multipurpose cadastre (stressing the representation of nearly all land information as attributes of parcel units) to embrace the concept of a multipurpose land information system (MPLIS) (Ventura 1991). An MPLIS incorporates various "spatially registered layers of institutionally independent data," including parcels as a primary spatial layer (Dueker and DeLacey 1990, 488–489).

As emphasized by Klosterman (1997, 52), "GIS alone cannot serve all the needs of planning," because it currently lacks support of "analytical and design functions that incorporate goals, objectives, costs, and benefits." Departments responsible for designing a land supply monitoring system must decide on an appropriate technology for each aspect of the monitoring process, considering GIS as but one among an array of tools available (e.g., remote sensing, spreadsheet models, database management software, automated permit tracking systems). Our survey of the LSCM literature and current practice suggests

---

[2] If only a fraction of a zone at the suburban fringe is densely developed, data aggregated for that zone would provide scant information as to the location of that development. Hence clustering, scattering, and nesting patterns of land uses and development, and land use mixes, cannot be tracked well at this level of analysis.

that parcel-based GIS offers a unique set of advantages, especially when used within the context of "multiple methods" and a "flexible set of routines" for monitoring land (Vrana and Dueker 1996, 23).

Parcel-based GIS: Advantages and Uses

The specific technical and methodological advantages of GIS as a tool for planning analysis, particularly as it relates to land supply and capacity, can be usefully summarized within categories developed by Webster (1993), which encompass (1) data organization and management, (2) visualization, and (3) spatial analysis. Each is discussed in turn in the following paragraphs, with examples illustrating parcel-based applications.

Data Organization and Management   GIS facilitates the management of both spatial and attribute data. It offers a set of specialized tools for spatial data creation, feature editing, attribute data retrieval, and data destruction (Webster 1993). These GIS functions are crucial to managing parcel data, which comprise the lines and polygons that represent parcel boundaries and areas, as well as the unique identifiers that link them to attribute records. Parcels are notably dynamic entities, especially in urbanizing contexts where frequent changes result from boundary adjustment, subdivision, and land assembly. A major function of PBGIS is to maintain ongoing consistency of the parcel base map with land records.

GIS also facilitates linkages between disparate land data attributes through the use of common geographic references, such as parcel identification numbers (PINs), standardized street addresses, xy coordinates, and zonal identifiers, thus utilizing what Webster (1993, 722) describes as "the facility [of GIS] for storing data from diverse sources as multiple attributes of geographic space referenced by some common form of geocode." The use of parcel identifiers by PBGIS to integrate data sets is not altogether new. Nonspatial land information systems have long used parcel identifiers to link tabular data. What PBGIS does offer, however, is the efficient storage and retrieval of data spatially referenced to the parcel, with the benefit of using the parcel map itself as an "interactive index" for retrieval, query, and flexible aggregation of the land data for various LSCM operations (Webster 1993). For example, specific land records can be retrieved by selecting parcels on the map display, and, in turn, parcels satisfying specific queries can be highlighted, mapped, and analyzed separately.

Visualization   The ability of GIS to display the characteristics of land in cartographic space (both on screen and through production of paper maps) allows for the identification of spatial patterns, locations, and trends that may not be apparent otherwise and "may or may not be supported by more rigorous statistical descriptions of revealed relationships" (Webster 1993, 721). PBGIS generally increases the precision and resolution with which land supply data can be visualized—especially elements of urban form; patterns of use, owner-ship, and valuation; specific sites where administrative actions apply; and market indicators of development potential.

The importance of visualization to the process of GIS analysis deserves emphasis. For example, in analyzing land supply and capacity, the observed clustering of parcels with common characteristics and within particular sub-areas commonly suggests additional variables to explore, or targeted public actions to test. In addition, fine-grained patterns evident in the display of parcel data often appear to correlate spatially with other types of features, such as roads or utilities infrastructure, indicating interactive phenomena. PBGISs also allow the visualization of alternative solutions to common plan-ning problems (such as proposed zoning changes, growth boundaries, regula-tory overlays). They allow planners and decision makers to evaluate the differ-ent factors that affect land development capacity, and to insert additional data variables or alter analysis assumptions to test the impacts of proposed investments or policy changes. The sensitivity of development capacity "mod-els" may also be tested readily with GIS, generating both mapped and statisti-cal output.

Just as significant as the role of visualization in analysis is its role in commu-nicating detailed geographic information about land supply and policies shap-ing it to decision makers, to different branches of local governments, and to the public. For example, serial maps showing the shifting locations of building permits over time can help policymakers understand the effectiveness of a growth boundary much more readily than nonmapped data.

Spatial Analysis Spatial analysis refers to the various GIS tools and func-tions that "[generate] new information based on explicit spatial processing" performed on either single or multiple layers of data (Webster 1993, 724). The following types of spatial analyses may be applied using parcel coverages, especially in conjunction with other spatial layers, to analyze and evaluate land supply and capacity:

- *Spatial overlay* allows various data types (e.g., parcels, environmental features, regulatory zones, and infrastructure elements) to be related to each other through cartographic superimposition of distinct layers. Parcel records may be related to data represented by nonparcel polygon features (e.g., environmental constraints, regulatory zones) through spatial overlay with these other layers. Parcels may also be aggregated to analysis zones, selected, summarized, or otherwise manipulated based on spatial correla-tion. In addition, the spatial and attribute data of multiple layers may be combined through "true" polygon overlay (e.g., union and intersect GIS functions) to create composite data layers (e.g., vacant land polygons with and without wetlands present, subparcel areas with differing zone designations). As a related spatial analysis function, point-in-polygon overlays may also be used to select parcel-point features according to shared characteristics or to aggregate them to various subareas for further analysis.

• *Buffer analysis* allows for the identification of features that lie within or beyond a specified proximity to other features. Such operations can increase the automation of site suitability screening over large areas. Specific considerations may include determinations of accessibility based on distance to transportation infrastructure, identification of serviced land based on distance to utilities infrastructure, and delineation of environmentally sensitive areas based on distance from critical natural features.

• *Topology* is the data model describing the spatial relationships (or connectivity) between features within a single GIS data layer. Topological relationships, such as adjacency, may also yield new information that is relevant to land supply analysis. Examples of such analyses include identifying opportunities for land assembly, assessing the impact of adjacent land uses, and calculating road frontage and access patterns to parcels or clusters of parcels.

• *Raster-based analyses* involve the use of GIS data in grid cell (raster) format. Recent advances in GIS technology allow for the ready conversion of vector data (represented as points, lines, or polygons) to raster format and back again. This opens up possibilities for analyses and calculations both within and between data sets. Large-area calculations based on the overlay of highly detailed data, such as parcel data, are rendered more feasible in raster format than they would be as vector data. Operations based on the distances between features may be carried out, such as neighborhood functions, which calculate new values for cells based on functions that reference statistical averages of the values of nearby cells. Land and improvement value, as well as various environmental factors, are examples of variables that may be evaluated with this technique.

The previously described benefits of parcel-based GIS for land monitoring match many of the specific needs of local government planners. The determination of whether and how to use GIS, however, typically hinges on the resources of individual jurisdictions—staffing, data availability, hardware and software—as well as on the overall stage of GIS development. In a recent study of land records modernization in three states, Tulloch et al. (1996) identified six stages of GIS/LIS evolution: (1) no modernization (with only tabular data computerized but no spatial reference), (2) system initiation, (3) database development, (4) record keeping, (5) analysis, and (6) democratization. The current state of PBGIS for land supply monitoring suggests that the field is at stage 2 or 3, with relatively few jurisdictions venturing extensively into or beyond stage 4 or 5 to employ the unique spatial analysis capabilities of GIS. Our survey of jurisdictions showed that the most common use of PBGIS, at or beyond a basic stage of database development, was for display and data management rather than for spatial analysis. However, the increasing breadth and depth of GIS implementation within local governments suggests that as further "evolution" occurs, PBGIS will soon expand in its application for LSCM.

CHALLENGES

The current state of LSCM practice, together with the advances in land information technology, points to several challenges for the future. Six such challenges emerge as particularly pressing.

Challenge 1: Establishing Definitions and Methods for LSCM That Are Credible and Relevant to Both Long-Term Public and Short-Term Private Interests

Future LSCM systems will have to satisfy both the new mandates for land monitoring to guide the implementation of growth policy and the various concerns expressed by the private sector. Success in this divided arena will entail developing approaches to LSCM that are more ongoing and methodologically nuanced than those currently in use. Monitoring will have to account not just for the long-term supply of land within broadly designated areas, but also for (1) the medium-term supply of land as it relates to adequate levels of urban infrastructure and services and (2) the short-term supply of buildable house lots and sites for multifamily, mixed-use, and commercial development.

PBGIS will likely become a useful tool to meet these requirements. Many of the parcel-level data now being collected for land records, assessment, and other public functions serve the day-to-day needs of administration *and* make possible an ongoing assessment of short- and medium-term land supply and capacity. Moreover, these data may be used (in aggregated or disaggregated form) for periodic assessments of the long-term prospects of the land supply in relation to both demand forecasts and the goals and objectives of urban growth policy. In this way, PBGISs promise to integrate the data requirements of both comprehensive planning and land management. They offer a means to "navigate" between long-term regional plan making and short- to medium-term local plan implementation and thus provide an information "bridge," functionally connecting the two realms of planning.

Challenge 2: Developing LSCM Methods That Address the Full Range of Buildable Lands and Evaluate Their Development Capacity Within a Detailed Geographic Context

Approaches to LSCM that have relied on relatively coarse data representations and analyses have been adequate to assess supply and capacity for large jurisdictions and entire regions. However, they lack the precision necessary to address many new growth management planning requirements.

Recent attempts to concentrate growth within small subregional areas, such as "Functional Design Areas" in Portland, Oregon, "Urban Villages" in Seattle, Washington, and within metropolitan cores more generally, require LSCM to address the precise locational characteristics of supply and capacity. Further, increasingly prevalent state-mandated concurrency between new land development and infrastructure provision, along with federal requirements for metropolitan area coordination of land use and transportation planning, suggest that land monitoring will have to be sensitive to adjacencies between land uses and development types and to relationships between serviced areas, available land supplies, and patterns of new and existing development.

Several factors have fueled a gradual reorientation of LSCM from an almost exclusive focus on vacant land at the urban fringe toward assessment of opportunities for infill and redevelopment. First, political resistance to significantly expanding urban growth areas continues to hold, despite strong demand and considerable pressure from the private sector. Second, urban revitalization, particularly in the area of redevelopment of underutilized or abandoned lands, is receiving renewed and overdue attention within growth management. These trends further imply a need to entertain considerations that are relatively new to urban land use planning, such as land use conversion, mixed land uses, brownfields redevelopment, the use of publicly owned lands, nontraditional approaches to private land development, and the underutilization of lands zoned for both residential and commercial uses.

LSCM tools and techniques can address these considerations in several ways. First, ongoing monitoring activities can capture data on development as it occurs, including physical and functional types, densities, locations, multiple uses, and preexisting uses. Second, supply analyses can identify underutilized sites or subareas for targeted public actions to encourage infill and redevelopment. Assessing the supply and capacity of public and institutional lands has gained in importance, and some cities have already begun to use PBGIS to inventory and craft redevelopment strategies for tax delinquent and otherwise surplus publicly owned properties. Third, land capacity analysis provides a "model" to test the potential impacts of areawide regulations to open up supplies of land for desired forms and concentrations of development.[3] Further, successful application of LSCM to support redevelopment and mixed-use projects should integrate systematic tracking of the dynamics of the land market with exploratory approaches to establishing appropriate incentives for the private sector. It should be noted that site-level data collection and analysis are essential to address infill and redevelopment and should encompass regulatory limits, market feasibility, and environmental constraints.

---

[3] Chapter 8 discusses the potential role of parcel-level dynamic urban simulation models to test the impacts of land use policies and capital facilities plans on the development of buildable land supplies.

The list of LSCM "tasks" addressing the potential of already urbanized and largely serviced areas to accommodate future growth may be quite daunting for local jurisdictions with limited resources for land monitoring. However, it should be clear that these comprise some of the most promising areas of land use analysis—particularly as supported by PBGIS—to promote resource-conserving urban development in the future.

Challenge 3: Database Development, Maintenance, and Data Sharing

Information system design and database development continue to be dominant activities for local government GIS programs. Among the jurisdictions surveyed for this project, GIS data layer construction was a primary focus of recent efforts. Developed primarily for management purposes, GIS layers such as parcels, utilities infrastructure, and environmental conditions effectively provide a detailed information infrastructure for LSCM. As this infrastructure continues to develop in a decentralized and piecemeal fashion, both within local jurisdictions and across regions, the key challenge for land monitoring agencies is to leverage the capture and enhancement of, and access to, data adequate to suit the specific requirements of ongoing monitoring and analysis.

This challenge extends beyond efforts to develop individual databases and includes information systems coordination and data sharing among multiple departments, agencies, and jurisdictions. The ability of GIS to integrate spatial and georeferenced data of many different types and from many different sources makes data sharing both feasible and desirable to a degree that is relatively new within and among the parochial domains of local government. The next step is to establish formal data sharing arrangements that meet the needs of all participants. To be most effective, the formalization of such arrangements should transcend the institutional separations that exist between local and regional governance, between land management and long-range planning functions, and between public and private sectors. This may be a difficult achievement, as it will entail the engagement of different institutional and professional "cultures" with divergent orientations toward data management and planning analysis. To complicate matters, new information technologies have introduced a new set of actors within local government—GIS and information systems specialists—whose perspective must be considered and with whom relationships must be established.

Improved coordination of GIS data sharing will specifically have to address issues ranging from spatial data standards (e.g., scale, spatial accuracy), to attribute standards (e.g., coding, classification schemes), to software and hardware compatibility. Efforts to "clean up" data—to improve their quality and comparability—will also be essential for effective and credible land monitoring. In addition, the establishment of technical and organizational protocols for data update will create the core procedures for ongoing land use change detection and land development monitoring. Automated updating of land supply data through links to transactional records of public and private actions (e.g., permit tracking, property sales records) holds potential for greatly improving the timeliness of LSCM.

Finally, networked and web-based GIS offer technological opportunities for improved sharing and updating. However, without concomitant progress in the organizational sphere, the situation within local government will continue to constitute what has been characterized as "unified access to a fragmented database" (Heikkila 1998). LSCM may provide a linchpin for more integrated GIS at the local and regional levels. As argued in Chapter 7, the expanding interest in, and even reliance on, land supply and capacity information by multiple departments creates the basis for mutually beneficial data-sharing agreements.

Challenge 4: Accounting for Market Conditions

Echoing the observations of Godschalk et al. (1986), our survey of current practice found that LSCM persists in evaluating land supply and capacity with little regard for market forces and conditions. This is especially the case relative to the availability of land supplies that may be feasible for development over the short-to-medium term. (See "Commentary" to Chapter 3 by Rolfe.)

Increasingly, monitoring agencies are pressured to adopt new approaches to incorporating market considerations. These may entail collecting and analyzing new types of data. Multiple Listing Service files, surveys of landowners' intentions, and monitoring of "pipeline" development (particularly platted house lots) can provide data from which short-term land supplies may be tracked with greater accuracy. (Current practice reveals uneven advances in this area, with the most notable improvements occurring in monitoring pipeline development.)

The supply of land *available* for development or redevelopment, however, can change considerably over time, depending on actual and perceived demand for land and space relative to supply. Land price may provide a barometer of this dynamic and constitute a more direct means to monitor land supply relative to demand than data on the land itself. However, linking price increases to specific land use controls may be prohibitively difficult, particularly at the regional level where frequent wide swings in demand and the relative inelasticity of the development industry to respond to these swings over the short-term will significantly influence price changes. Nonetheless, the comparison over time of subregional variations in price for both improved and unimproved lands may be useful to alert local jurisdictions to bottlenecks in supply.

Finally, improved demand forecasting, particularly at the subregional level, can enhance the assessment of supply adequacy by local governments. Such forecasting should be sensitive to local supply constraints, as well as to the dynamic relationship between producers and consumers in the real estate marketplace. (See Chapter 8 for a discussion of the role of urban simulation models in improving demand forecasts.)

Challenge 5: Addressing Environmental Impacts and Constraints

LSCM typically treats environmental constraints on development as a static set of regulation-defined conditions that effectively constrain the supply of buildable land. These conditions include floodplains, wetlands, steep slopes, riparian corridors, and others, depending on local situations and policies. Such treatment, however, falls short of addressing both the cumulative and dynamic aspects of environmental conditions in urbanizing areas.

The responsiveness of LSCM to environmental factors can be enhanced in several ways. In projecting future land development, land monitoring can provide detailed information on the location of new development relative to sensitive areas and can potentially serve to quantify the cumulative impacts of development on these areas. Less site-specific impacts of urban development can also be tracked and analyzed in coordination with LSCM, to include mapping of pollution "sheds" around future industrial and transportation concentrations, monitoring land development within watersheds (and related potential disruptions of natural drainage patterns), and providing data to estimate the cumulative impacts of development on land cover and soil permeability.

In addition, to the extent that efforts to concentrate urban growth may intensify environmental conflicts between adjacent or proximate land uses, land monitoring can capture data relative to the environmental degradation resulting from urban concentrations. As urbanization may itself engender environmental effects, interactions between the urban growth process and the environment should be considered in the assessment of land supply and capacity over the long term. Environmental factors are likely to have an increasingly important impact on the location of the demand for land, thereby affecting its price, the resulting patterns of land use, and even the overall potential for growth within urbanizing regions. Research is certainly needed in this area, yet it seems clear that LSCM, performed in concert with environmental planning and modeling, can be used to project the scope of future regulations within different scenarios. It may thus help to design future regulations and incentives in such a way as to direct new development to areas where ecological and environmental impacts are least severe. (See "Commentary" to Chapter 6 by Marina Alberti for further discussion of environmental factors.)

Challenge 6: Securing Political and Financial Support for Ongoing Land Supply Monitoring That Meets the Requirements of Challenges 1 Through 5

It is fairly apparent that the activities covered in the five preceding challenges demand significant financial and institutional backing. The development of new LSCM methods requires skilled staff with adequate amounts of time and institutional support to "experiment" with such approaches to land monitoring and analysis. Developing a comprehensive database infrastructure for ongoing LSCM, especially if it is heavily reliant on GIS, also entails considerable costs. However, the success of coordinated data sharing may offset these costs (see Chapter 7). Finally, efforts to broaden the scope of LSCM to deal effectively with the dynamic aspects of land markets and urban ecology will likely require further research in impact analysis and modeling. Because such work is generally beyond the reach of the public sector, increased funding for academic research that explicitly addresses LSCM is needed to expand the field.

OPPORTUNITIES

Opportunities abound for land monitoring to become an effective tool in supporting efforts to improve urban land development patterns and practices. Two separate forces appear to converge in such a way as to broaden future applications of LSCM to a wide range of jurisdictions, as well as to improve its relevance to land development practices. One is the growing trend nationally toward managing the impacts of future urban development through growth-conscious planning, and the other is the rapid development of GIS and other information technology to serve both planning and land management functions. These forces together suggest new opportunities for LSCM.

The number of metropolitan regions considering or implementing strategies for managing urban development is growing across the country. As evidence of the strength and national scope of growth-management, the November 1998 elections saw citizens in New Jersey and Arizona support initiatives to fund large-scale open space acquisition programs, and Tennessee recently joined the ranks of growth management states. Recent trends also appear to herald a growth-conscious planning that stresses the fiscal imperatives of containing new development (Peirce 1997; Rosenberg et al. 1995). The State of Maryland Smart Growth legislation has been featured as a model in this arena. At the national level, a further sign of tightening controls on urban development is the continuation of federal legislation requiring the integration of land use and transportation policy—namely, the Intermodal Surface Transportation Efficiency Act (ISTEA) and the Transportation Equity Act for the 21st Century (TEA-21). Such legislation requires not only that linkages be made between urban development, infrastructure, and environment, but also that the interactions between them be assessed. This introduces incentives for local and regional governments to adopt more sophisticated monitoring and modeling tools.

The implementation of growth management regulations creates a need for improvements in the practice of LSCM. To the degree that new and intensified controls on growth restrict land supply relative to demand, competition for land increases. In response, the private sector calls for more frequent monitoring of land supply—at a high level of detail and addressing both short- and long-term supply—which relates to the interests of the real estate and land development industries. At the same time, public concerns about the potential impact of growth controls on housing affordability and economic vitality imply similar land monitoring requirements. Furthermore, planners require reliable methods to measure plan effectiveness, especially as an input to assessing the costs and benefits of regulations. Such methods extend beyond assessing land supply within urban growth areas to consider transfer of development rights, concurrency, shoreline protection, and other regulatory tools to manage growth.

Overall, a continued spread of growth-conscious planning is to be expected as increases in both population and standard of living place greater demand on the urban land supply and infrastructure and strain environmental systems. From both public- and private-sector perspectives, the pool of lands available for future urban development is visibly shrinking and will continue to do so. In this sense, LSCM is bound to become more central to planning and to require increasingly sophisticated methods in the future. Preparing for this eventuality, the American Planning Association's current Growing Smart program has included a section on land supply monitoring systems (with a model legislative mandate for LSCM) in its new guidebook for the revision of state planning enabling legislation. The guidebook refers to land supply monitoring systems as "the missing link in many critical local development decisions made by public policymakers and development interests" (American Planning Association 1998, 7–143).

In parallel to growth-conscious planning, the consolidation of GIS as a decision and management support system for local government offers the promise of new forms of LSCM that meet the challenges outlined earlier. As argued, the rapid advances in GIS technology, accompanied by an explosion in the availability of spatial and spatially referenced data, are altering the relationship between land management and long-range planning. With this technology, parcel-specific data, including traditional land records, land transactions and permitting, existing and planned land use, and infrastructure provision, can now be effectively managed and evaluated to serve both functions. GIS and networked information systems enable the requisite data sharing, potentially contributing significantly to reducing the costs of land monitoring. With much of the available data at a high level of resolution (particularly for parcels), they can be flexibly aggregated to other units of analysis and display. PBGIS thus offers the potential for meeting the planning and management needs of multiple public sector users, with increased relevance to the private sector as well.

In this sense, PBGIS is likely to contribute to opening up a "market" for LSCM. Improved multipurpose land information databases support jurisdictions in developing complex and diverse applications of both land monitoring techniques and land supply and capacity information. These potential trends augur a "democratization" (Tulloch et al. 1996) of land information systems at the local level, characterized by increasingly decentralized access to and use of GIS tools and data.

Technological and organizational changes will greatly change the nature of the field and the actors and decision makers within it—so much so, in fact, that planning may not remain at the center of LSCM. The future depends very much on the success of efforts to coordinate GIS for effective land monitoring. Such coordination will entail, in part, the adoption of standards for both data and methods, as well as the integration of LSCM into multiple functions of local government. Two contrasting opinions prevail regarding what this future may be like. On one hand, Innes and Simpson (1993) suggest that there is generally little interest in developing truly comprehensive systems, largely because of parochial specifications for what a GIS should be. On the other hand, recent developments in both planning practice and planning applications of GIS may represent, as Heikkila (1998, 359) suggests, a "growing technological basis for collaboration coupled with a powerful imperative to collaborate." The requirements for and benefits of implementing a comprehensive land supply and capacity monitoring system, especially as pursued within a framework of mandated growth management planning, may indeed bolster that imperative.

REFERENCES

American Planning Association. 1998. *Growing Smart legislative guidebook* Phase II. Interim ed. Chicago: American Planning Association.

Bollens, Scott A. 1998. Land supply monitoring systems. In *The Growing Smart working papers*. Vol. 2. Chicago: American Planning Association.

Bollens, Scott A., and David R. Godschalk. 1987. Tracking land supply for growth management. *Journal of the American Planning Association* 3: 315–327.

Castle, Gilbert H. 1997. The bigger picture. *Business Geographics* 2: 16.

———, ed. 1993. *Profiting from a geographic information system*. Fort Collins, Colo.: GIS World Books.

De Grove, John M. 1992. *Planning and growth management in the states*. Cambridge, Mass.: Lincoln Institute of Land Policy.

Diamond, Henry L., and Patrick F. Noonan, eds. 1996. *Land use in America*. Washington, D.C.: Island Press.

Donahue, James G. 1994. Cadastral mapping for GIS/LIS. *ACSM/ASPRS international proceedings page.* <http://wwwsgi.ursus.maine.edu/gisweb/spatdb/acsm/ac94114.html> (January 22, 1999).

Drummond, William. 1997. Telephone interview by author. December 12.

Dueker, Kenneth J., and P. Barton DeLacey. 1990. GIS in the land development planning process: Balancing the needs of land use planners and real estate developers. *Journal of the American Planning Association* 4: 483–491.

Enger, Susan C. 1992. *Issues in designating urban growth areas*. Parts I and II. Olympia, Wash.: State of Washington, Department of Community Development, Growth Management Division.

Godschalk, David R., Scott A. Bollens, John S. Hekman, and Mike E. Miles. 1986. *Land supply monitoring: A guide for improving public and private urban development decisions.* Boston: Oelgeschlager, Gunn & Hain, in association with the Lincoln Institute of Land Policy.

Heikkila, Eric J. 1998. GIS is dead; Long live GIS! *Journal of the American Planning Association* 3: 350–360.

Holmberg, S. C. 1994. Geoinformatics for urban and regional planning. *Environment and planning B: Planning and Design* 21: 5–19.

Huxhold, William E. 1991. *An introduction to urban geographic information systems.* New York: Oxford University Press.

———. 1997. Interview by author. October 11.

Innes, Judith E., and David M. Simpson. 1993. Implementing GIS for planning: Lessons from the history of technological innovation. *Journal of the American Planning Association* 2: 230–236.

Klosterman, Richard. 1997. Planning support systems: A new perspective on computer-aided planning. *Journal of Planning Education and Research* 17: 45–54.

Kollin, Cheryl, Lisa Warnecke, Winifred Lyday, and Jeff Beattle. 1998. Growth surge: Nationwide survey reveals GIS soaring in local governments. *GeoInfo Systems* 2: 25–30.

Korte, George. 1997. *The GIS book.* Santa Fe: OnWord Press.

Masser, I. 1984. *Strategic monitoring for urban planning in developing countries.* TRP-51, Department of Town and Regional Planning, University of Sheffield, Sheffield.

Nedovic-Budic, Zorica D. 1993. GIS use among southeastern local governments: 1990/1991 mail survey results. *URISA Journal* 5(2):4–17.

Orman, Larry. 1997. Computer mapping: New frontier for land use and real estate professionals. *Lusk Review* 1: 51–61.

Peirce, Neil. 1997. Maryland's "Smart Growth" law: A national model? *Washington Post Writers Group page.* <http://www.clearlake.ibm.com/Alliance/newstuff/peirce/peirce_042097.html> (October 21, 1997).

Public Technology, Inc., Urban Consortium, and International City Management Association. 1991. *Local government guide to geographic information systems: Planning and implementation.* Washington D.C.: Public Technology, Inc., Urban Consortium, and International City Management Association.

Rosenberg, D., et al. 1995. *Beyond sprawl: New patterns of growth to fit the new California.* Report sponsored by Bank of America with California Resources Agency, Greenbelt Alliance, and Low Income Housing Fund.

Stein, Jay M., ed. 1993. *Growth management: The planning challenge of the 1990s.* Newbury Park, Calif.: Sage.

Thrall, Grant. 1997. Telephone interview by author. November 19.

Tulloch, David L., Bernard J. Niemann, Earl F. Epstein, W. Frederick Limp, and Shelby Johnson. 1996. Comparative study of multipurpose land information systems development in Arkansas; Ohio, and Wisconsin. In *GIS/LIS '96: Annual conference and exposition proceedings.* Bethesda, Md.: American Society for Photogrammetry and Remote Sensing, 128–141.

Ventura, Stephen J. 1991. *Implementation of land information systems in local government—Toward land records modernization in Wisconsin.* Madison: Wisconsin State Cartographer's Office.

Vrana, Ric, and Kenneth J. Dueker. 1996. LUCAM: Tracking land use compliance and monitoring at Portland Metro. Report to Metro, Portland, Oregon. Portland: Center for Urban Studies, Portland State University.

Webster, C. J. 1993. GIS and the scientific inputs to urban planning. Part 1: Description. *Environment and Planning B: Planning and Design* 20: 709–728.

2

Elements of a General Framework for Land Supply and Capacity Monitoring

Michael Hubner and Anne Vernez Moudon

Land supply and capacity monitoring (LSCM) encompasses all the activities necessary to track the current use of land and to assess its potential future use. As a descriptive concept, land "use" has multiple dimensions, which include *activities* occurring in specific *locations,* the *structures* housing those activities, and the *characteristics of the land* itself (Kivell 1993; American Planning Association 1999). Figure 2.1 illustrates the different components of land use and the relationships between them, particularly as they are considered for the purposes of LSCM. At one end of the spectrum are land, natural systems, and environmental regulation, and, at the other end, people, activities, and markets for land and developed real estate. These elements converge to determine how and where people and activities are housed in structures and other built elements on land that is supported by a common infrastructure of iservices. Figure 2.1 presents a simplified characterization of land use, which echoes the structure of LSCM systems that effectively relate "developed and developable land to infrastructure availability, environmental quality and constraints, and market trends" (Kaiser et al. 1995, 199).

Monitoring Land Supply with Geographic Information Systems, edited by Anne Vernez Moudon and Michael Hubner ISBN 0 471371673 © 2000 John Wiley & Sons, Inc.

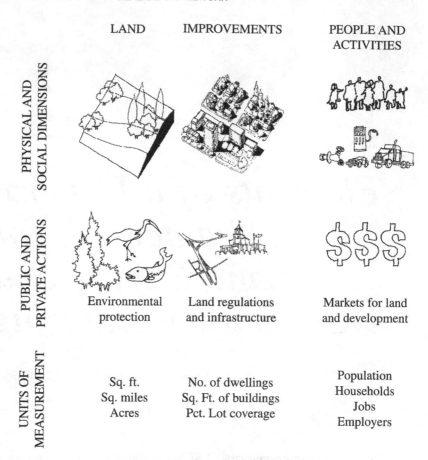

FIGURE 2.1   *Components of land use*

Specific approaches to monitoring and analyzing land supply and capacity all share a common purpose—to track changes related to land use and use potential—yet they also vary considerably. Generalized LSCM models have been proposed (Godschalk et al. 1986; Enger 1992; Dearborn and Gyri 1993), but no consistent approach or methodology has emerged in practice. Most applied methods are in-house creations. Cities, counties, metropolitan planning organizations (MPOs), and even states have established their own LSCM methods and assumptions that fit their distinct approaches to managing land resources. As may be expected, the most sophisticated land supply monitoring systems are maintained by those jurisdictions where growth management laws support, and in some cases mandate, such monitoring activities. However, additional local planning priorities, ranging from watershed protection to natural hazards mitigation to economic revitalization, have also driven efforts to monitor land and its capacity for developed use.

This chapter places the different pieces of the many existing individual LSCM approaches into one general framework. The elements of this framework are (1) the development of a land supply database, (2) an inventory of the buildable land supply, (3) the analysis of land capacity, and (4) the application of land supply and capacity information to urban and regional planning. The chapter proceeds with descriptions of what each element consists of, with specific examples of applications in practice and references to other chapters and appendixes where appropriate. Because land monitoring practices remain varied, homegrown, and ever-changing, considerable attention is given to the technical and methodological issues that continue to make LSCM a challenging endeavor. The reader will find several checklists of issues and considerations to be taken into account within each of the framework's four elements. First, however, some definitional groundwork is laid, which is necessary to discuss the range of issues.

DEFINITIONS

An essential step in elaborating a framework for land monitoring is to distinguish between *land supply* and *land capacity*. These terms are neither clearly nor consistently defined in the literature or in practice. In a broad, comprehensive sense, land *supply* describes the entire land base within a jurisdiction or urban region. Vacant lands, both at the fringe and within urbanized areas, may accommodate new development. However, it is important to consider, as well, that "developed" lands undergo continuous processes of alteration, expansion, and redevelopment as they adapt to new uses and users. Land *capacity* refers to the types and amounts of development (i.e., buildings and other improvements) and activities (e.g., households and employment) the land supply is capable of supporting. The schematic presented in Figure 2.1 mirrors the distinction established here between supply (as related to land) and capacity (as related to development). From a land supply monitoring perspective, it is important to recognize that a given piece of land can accommodate a broad range of types and amounts of development, depending on how the land is regulated (including how flexible or dynamic the regulations are over time), how desirable its location is for specific uses, and how much competition exists for the use of that land.

A second "cut" in distinguishing between supply and capacity identifies that part of the land supply that is available and capable of supporting *new additional* development along with the additional households and employees that may be accommodated within it. To this end are posited the following definitions:

- *Buildable land supply* is that land on which additional or new development can occur within regulatory, physical, and market-imposed limits. Buildable land supply is expressed as an amount of land.

- *Development capacity* is the amount of additional and new development that can occur on land identified as "buildable." Development capacity is expressed as a quantity of built space, such as dwelling units or building square footage, or may alternatively refer to households or employees.

Further, three general "levels" or types of measures of both buildable land supply and development capacity may be distinguished. Here these levels are referred to as "maximum," "adjusted," and "potential projected" supply and capacity (Figure 2.2).

Maximum supply and capacity establish, respectively, the greatest amount of land that can be developed and the greatest amount of development that can occur within the parameters set by zoning and other regulations, infrastructure, and prohibitive environmental constraints. This has also been referred to as "theoretical" or "gross" supply or capacity, or simply "build-out."

Adjusted supply or capacity encompasses various reductions to the maximum estimate to account for observed or anticipated factors that mitigate against full build-out. Adjustments may reflect land markets and the decisions of individual landowners (that limit the availability of land) as well as land development and construction practices (that determine the type and density of construction). In addition, adjusted supply may account for various uncertainties in public planning and management, such as the timing of extensions of infrastructure and public services or anticipated changes (both large and small) to land use plans and associated regulations. Adjusted supply or capacity is also referred to as "available" or "net" supply or capacity.

Establishing maximum supply or capacity can be done at a technical level, because it relies on a set of established policies and generally assumes that full build-out will occur. Estimating adjusted supply or capacity, however, is more complex and involves making multiple judgments within a potentially contentious political environment. Decisions must be made about what assumptions and criteria will govern the analyses of which lands are *likely* to

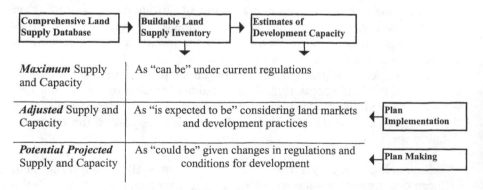

Comprehensive Land Supply Database	Buildable Land Supply Inventory	Estimates of Development Capacity	
Maximum Supply and Capacity	As "can be" under current regulations		
Adjusted Supply and Capacity	As "is expected to be" considering land markets and development practices		Plan Implementation
Potential Projected Supply and Capacity	As "could be" given changes in regulations and conditions for development		Plan Making

FIGURE 2.2 *Levels of supply and capacity*

be developed and how they will be developed. This involves consideration of multiple factors (e.g., site constraints, consumer choices, and land ownership) that influence development "outcomes." In planning terms, these outcomes include land use mix; the density, configuration, and timing of development; redevelopment and infill versus greenfields development; and residential and commercial under-build.

As a third level or type of analysis, *potential projected* supply or capacity entails moving from an essentially static account, given a set of current regulations and development conditions, to a dynamic assessment of land, its regulation, and potential development in the future. This includes testing future policy alternatives, ranging from modest assessments of the impacts of local zoning changes to full-scale analyses of regional land use plans. It also includes efforts to account explicitly for anticipated or potential shifts over the medium to long term related to land development and utilization, household size, employment mix, continuation of current or planned zoning regimes, and other important variables. This work may also entail analyzing supply and capacity for multiple time periods, either overlapping (short-, medium-, and long-term) or sequential (e.g., five-year increments). Designing an approach to assessing potential projected supply and capacity quickly means leaving the technical arena to engage the politics of long-term strategic planning. Dealing with a range of potential conditions, however they are addressed, increases reliance on professional (and political) judgment to guide data collection and analysis processes.

FRAMEWORK ELEMENTS

The preceding definitions lay the groundwork for discussing the four primary elements, or tasks, of land supply monitoring and capacity monitoring:

1. Developing a comprehensive land supply database
2. Conducting an inventory of the buildable land supply
3. Estimating development capacity
4. Applying land supply and capacity information to plan-making and plan implementation processes

Although the four elements are conceptually and functionally distinct, they are also related dynamically in practice, particularly as they influence data requirements and methods of analysis and as they fit within the overall design and utilization of the land supply monitoring system. The rest of this chapter examines these elements and highlights some of the major issues that most directly touch on them. The presentation is intentionally general; the reader may refer to Appendix A, an analysis of survey findings, or Appendix B, representative case summaries, for detailed examples from practice.

Element 1: Comprehensive Land Supply Database

A first and most basic element in the process of monitoring land supply is the capture, representation, organization, and maintenance of a range of data related to land and land development. These are the components of a comprehensive inventory of the entire land supply (broadly defined), its use characteristics, and other features of the landscape, both physical and regulatory, that shape the use potential of land.

It is important to consider the development of a comprehensive land supply database as separate from the buildable land supply inventory and analysis of development capacity. Several reasons underlie the distinction. First, as noted previously, the data that are potentially useful to land supply monitoring are at once dependent on, and useful to, other general planning activities, as well as to a range of other distinct local government functions. In addition, because approaches to monitoring and analyzing land supply can, and often do, change over time, it is useful to emphasize the maintenance of an independent inventory of land supply data. The database must be as complete and comprehensive as possible (given budgetary limits and the ability to anticipate information needs), rather than narrowly circumscribed to fit often ephemeral definitions of supply and capacity.

Generally, elements of a land supply database include the following data types:

- Existing and planned land uses
- Zoning and related regulatory overlays
- Other regulations that impose limits on density and use (e.g., subdivision ordinances, building codes, site design codes, and parking requirements)
- Census and other demographic data
- Data derived from remote sensing images (as well as the image data themselves)
- Land ownership information
- Assessed valuation (of both land and improvements) and taxation status
- Age of improvements
- Environmental conditions and natural features, including shorelines, rivers and streams, topography, wetlands, steep slopes, soils, and natural areas
- Existing and planned infrastructure (including service levels and capacities)
- Development "in the pipeline," particularly as derived from permits (both for land subdivision and building construction)
- Market-related data (including Multiple Listing Service files, records of sales transactions, and indicators of land availability)

Godschalk et al. (1986) stressed the importance of collecting market-related data as part of a land supply monitoring system, particularly data related to land availability and the economic feasibility of development. Others have followed with similar recommendations (Enger 1992; Bollens 1998). So far, however, there are few land supply databases that systematically combine such market data with physical or regulatory indicators of developability. Continuing barriers to the generation of such databases include proprietary interests in such data (both private and public) and lack of compatibility between public- and private-sector land information systems.

Ideally, nearly all of the data types included in the preceding list would be located geographically as features or attributes of geographic information system (GIS) layers. Efforts should also be made to capture and maintain longitudinal data, especially records of land use change and development activity (see Chapter 6).

The following considerations are important in the design and development of a comprehensive land supply database:

- Quality of the data, including accuracy, precision, currency, verifiability, clarity, quantifiability, freedom from bias, and availability and completeness of metadata (Dale and McLaughlin 1988)
- Comparability and compatibility of data types and formats (especially GIS coverages) both over time and across jurisdictions and agencies
- Geographic extent and completeness of data collection to encompass both present and future locations of urban development
- Spatial units for which data are collected and represented in the database, particularly as they determine allowable levels of disaggregation for supply and capacity analyses
- Representation of heterogeneous characteristics within single geographic features, such as multiple uses on one parcel or multiple owners or ownership types within one land use area (See Chapters 3 and 6 for further discussion of this issue.)
- Representations and models of temporal events and relationships, particularly as they inform analyses of trends and land use change processes

As discussed in detail in the following sections, the significance of specific data issues is driven by various policies and methodological requirements. However, decisions made by local governments about database development and land information system design are also driven by the balance of anticipated costs versus potential benefits. Data capture, including both new data and the conversion of existing data and maps to usable formats, typically represents 60 to 80 percent of system development costs, with parcels generally being the most costly data layer to produce (Public Technology, Inc. et al. 1991; Korte 1997). For this reason, jurisdictions usually undertake such work based on the findings of a needs assessment that links specific data types to the identified requirements of departmental activities. Our survey of GIS used within local government land monitoring systems revealed widespread data sharing between departments and (less commonly) between jurisdictions.

These practices may serve to mitigate many of the high costs associated with independent efforts to monitor land supply, especially at the resolution of the tax-lot parcel (see Chapter 7 for further discussion).

Element 2: Inventory of Buildable Land Supply

A buildable land supply inventory identifies lands where new or additional development is reasonably able to occur. The delineation of such a supply is geographically specific and is expressed in units of land, such as acres. Some inventories refer, as well, to numbers of house lots; this is, however, technically a capacity measure (see the next section). Ideally, the inventory assumes a specific time horizon for land development, thus implying expected conditions for both land availability and amounts of demand.

Typically, buildable lands are classified as one of three types:

- *Vacant:* Lands without any significant uses or improvements
- *Partially utilized:* Improved parcels that are sufficiently vacant to allow additional new development without replacing or altering existing structures (commonly referred to as having infill potential)
- *Underutilized:* Properties already developed over most of their area, but considered likely to undergo demolition and replacement or significant alteration of structures to yield more intense, more highly valued, or altogether different uses (commonly referred to as having redevelopment potential)

In order to identify buildable lands in each of these categories various measures and methods are commonly applied to perform an initial screen of the entire land supply, as follows:

- For vacant lands: air photo interpretation, assessor-assigned or in-house land use codes, improvement value (per land area or value)
- For partially utilized lands: aerial photo interpretation, ratio of improvement value to land value, development density (e.g., dwelling units per acre, floor area ratio, lot coverage) compared with that allowed by zoning and other regulations
- For underutilized lands: ratio of improvement value to land value, ratio of improvement value to land area, comparison of valuation ratios with those of nearby parcels

Typically, the three types of buildable lands are further classified by *zoning* or *generalized land use plan designation*—residential, commercial, industrial, or, less commonly, mixed use. Buildable lands may also be classified according to the level at which each type of land is or will be provided with *urban services and infrastructure* adequate to support planned uses.

The following types of land are usually excluded from the buildable land supply:

- Lands constrained by environmental or physical factors that prohibit development (largely determined by prevailing regulations limiting the use of these areas)
- Public or other lands whose ownership status likely precludes private development (e.g., churches, schools, cemeteries, and land with public easements, rights-of-way, or other restrictions to ownership and development rights)
- Lands set aside for public purposes (e.g., streets, educational facilities, drainage, and open space). These exclusions typically apply to greenfields and entail either removing a set of dedicated sites from the inventory or applying land use ratios or standard discounts to deduct land required for public purposes.

Consideration of the maximum buildable supply entails identifying all vacant and partially developed lands that are not constrained by the factors thus far outlined. The adjusted supply typically reflects a consideration of land and development markets and practices, as well as a more discriminating assessment of adequate service levels, accessibility, potential redevelopment costs, and other factors where uncertainty is present. As noted, most buildable land supply inventories focus on vacant land, especially that planned for single-family residential development. Jurisdictions do not often consider the supply of small, partially developed (infill), or redevelopable parcels. This may be due to lack of data or to limited capabilities for analyzing the data. Frequently, as well, jurisdictions performing land supply and capacity monitoring (LSCM) perceive that demand pressures relative to supply are not strong enough to stimulate redevelopment or infill within already urbanized areas. In the absence of a mandate or specific framework for making more discriminating assessments of "developability" or "redevelopability," most jurisdictions continue to emphasize the development potential of greenfields and to rely on limited land records and land use categories to inventory land supply.

The following are the major issues related to conducting an inventory of buildable lands:

- Geographic extent—the *area* over which the inventory is made—has an impact on the long-range utility of the buildable land supply inventory as a baseline for future monitoring. Extent is especially important relative to consideration of areas not yet in urban use, but which may be annexed or developed with future urban expansion. Lands considered for potential annexation have been the focus of several monitoring programs. In Boulder, Colorado, the county planning department conducts regular land supply studies for areas to be annexed by municipal jurisdictions. Two examples in California, the city of Ontario and Marin County, also illustrate the utility of extending land monitoring beyond municipal borders to assess supply and capacity (in these cases into areas designated as "spheres of influence," delineated for long-term phased annexation). Another example is Washington's Buildable Lands legislation, which mandates the monitoring of land supply data both inside *and* outside of Urban Growth Boundaries (UGBs).

• Screening or analyzing the overall land supply in order to inventory buildable lands will likely involve a choice of spatial unit or units, and data aggregation. The parcel or the tax lot is increasingly the preferred unit of record and analysis, largely because of the central role that ownership plays in trading, improving, and regulating land. However, the buildable land supply may also be inventoried utilizing other units, such as homogeneous land use areas (e.g., polygons representing vacant lands in GIS coverages), zoning districts, analysis zones, or units based on the intersection of different units represented on multiple GIS layers (see Chapter 3 for a description of such an approach).

• Environmental and physical constraints to development should be assessed with accurate data and appropriate methods. In practice, such constraints are either identified through spatial overlay in a GIS or otherwise accounted for by applying percentage deductions based on estimates of the presence of constraining factors in an area. In the absence of strong and clear regulations governing the use of environmentally sensitive areas, it is difficult to relate physical constraints precisely to future development potential.

• Determining the developability of land (especially vacant land at the fringe) relative to the provision of adequate infrastructure and servicing may be problematic. First, private development may be difficult to predict if there is no clear policy framework linking infrastructure and development regulations (as in the case of adequate public facilities ordinances used by Montgomery County, Maryland, or concurrency requirements in the State of Florida). Similar problems arise if capital improvement plans (CIPs) are not consistent with land use plans. Second, it may be difficult to determine what constitutes "serviced" land, or what qualify as "adequate" levels of service for future development. Certainly, the presence or absence of major infrastructure elements (i.e., transportation, water, and sewer) is important, but analyses may have to consider other types of servicing that typically accompany residential development, to include schools, parks, public safety, and even private-sector "facilities" such as retail. So far, most land supply analyses have excluded these latter considerations. Third, a proper distinction must be made between macroserviced land (generally referring to proximity to major trunk-line infrastructure, such as sewer mains and arterial roads) and microserviced land (generally indicating direct provision of infrastructure and services to individual sites or parcels). Fourth, translating levels of infrastructure service capacity into discrete land areas or amounts presents significant difficulties, which include (1) determining the relationship between future development density or type and anticipated service demand; (2) dealing with the impact of variable or flexible levels of service or infrastructure provision within urban areas—for example, higher levels of congestion may be tolerated in the central city than in suburban settings; and (3) accounting for the cumulative impacts of development on the demand for multiple services, especially demand imposed by mixed-use or phased projects.

- Screening for buildable lands involves multiple criteria applied either serially or simultaneously through spatial overlay. This process, which essentially entails matching land uses to appropriate locations, mirrors standard methods of land suitability analysis. By extention, potential drawbacks of standard suitability screening methodologies also characterize common methods of land supply analyses. First, buildable lands screens do not usually account for potential interrelationships between factors that may constrain developability. Second, multiple suitability factors are not usually weighted to reflect their relative importance to development outcomes. Third, inadequate standards may be in place for what constitutes "good" and "bad" data. Finally, the impacts of "intangible" factors—those that may not be easily measured and quantified—are not typically considered in the assessment of buildability (Anderson 1987).

- Various approaches may be used to identify lands with redevelopment potential. As noted, a common method involves screening for sites based on the calculated ratio of assessed improvement to land value for each parcel. This analysis establishes a threshold ratio below which market pressure is presumed to induce owners to replace existing structures with more highly valued and revenue-producing land uses (e.g., conversion of single-family to multifamily residential or commercial use). The threshold value varies considerably by case; for example, the City of Seattle has used a 0.5 cutoff for commercial and mixed-use redevelopable sites, and Montgomery County a ratio of 1.0 for commercial lands. Other monitoring programs, such as Lane Council of Governments', have employed mixed screening approaches using an improvement to land value ratio along with other indicators. An overarching concern is that either simple ratios or composite indicators, applied evenly across large study areas, may not be valid in indicating redevelopment potential for specific subareas and submarkets. More subarea-specific analysis requires "an accurate sense of local market conditions, policies, and values," considerations that will be challenging to work into areawide monitoring approaches (Enger 1992, 13).

- The supply of platted single-family parcels in the development "pipeline" may constitute a significant proportion of the "supply" of buildable land for residential uses. These parcels should, however, properly represent capacity for single-family residential development—creation of house lots effectively sets the development capacity of such land. Land monitoring tools can be used to track the development of lots in large-scale subdivisions, especially in jurisdictions where a historical legacy of speculative subdivision represents a significant land management issue. Two prominent examples—Lee County, Florida, and Kansas City, Missouri—have used parcel-based geographic information systems (PBGIS) to track the market absorption and project completions within large subdivisions (as large as 300,000 lots in Lee County).

- Criteria for deducting land from the maximum supply to arrive at an adjusted supply estimate should reflect marketwide conditions. This may involve the application of a "market factor" to account for land kept off the market for a variety of reasons, including speculation, land banking, future expansion, personal use, and other vagaries of the market. Also used as a cushion against land and housing price inflation, the market factor is intended to allow adequate flexibility for developers and real estate investors to locate in areas favorable to the market (Easley 1992). The market factor is applied as a percentage deduction to the total buildable land supply or, more commonly, to aggregate development capacity (see the discussion in next section regarding a market factor applied to capacity).

Element 3: Estimation of Development Capacity

Development capacity analysis converts the buildable land supply into measures of development, quantities such as the number of dwelling units or the square feet of nonresidential space, that can be accommodated by the supply. At its most basic level, the conversion from units of land to units of development involves multiplying the area of developable land in different land use categories by the density allowed under, or implied by, regulations governing subdivision and building construction for those uses. This is typically done according to zoning, but may also follow limits implied by planned land use categories. Establishing maximum capacity simply considers the gap between existing use and density and the regulatory envelope for allowed uses and densities. An adjusted capacity reflects an expectation that development will occur at a level somewhat below the zoning envelope, depending on the presence and strength of a number of factors (discussed further in the following paragraphs). PBGIS can lend considerable geographic detail to the analysis of the determinants of both maximum and adjusted capacity.

The following are the major issues involved in the analysis of development capacity:

- An appropriate and explicit time frame for the analysis must be established. Typically, LSCM is performed for time horizons of 5 to 20 or more years. The analysis horizon (whether explicit or implicit) largely determines expected levels of demand and construction activity, usually corresponding to anticipated growth in households and employment. The length of the analysis horizon will also strongly influence assumptions about the relative strength of anticipated market forces.

- Capacity analysis may be conducted at various levels of disaggregation to determine the development potential for land within parcels, planning subareas, watersheds, neighborhoods, or areas defined by zoning or planned land use. The unit of analysis will largely determine the flexibility within which capacity estimates may be aggregated and applied to multiple purposes. In most cases, parcel specificity is the highest feasible level of disaggregation. A number of agencies surveyed assigned values to individual parcels indicating measures of developability and capacity.

- Standard land use ratios, based on observed national or regional development trends, are sometimes used to assess the capacity of vacant land at the fringe (greenfields) for future growth (Easley 1992). Such ratios vary significantly over time. Recent evidence reveals a broad trend for residential and commercial uses to increase relative to industrial and public uses (Harris 1992). Within specific regional or local contexts, such ratios should be used with caution, especially for estimating capacity over the long term (15 to 25 years).

- Site-level conditions that do not per se prohibit development (i.e., as spelled out in regulations) may still restrict its economic feasibility (Moore 1997; Dueker 1998). Variability in demand, either by location or over time, will alter the degree of effective constraint on individual sites, thus adding complexity to analyses of the influence of these factors.

- Mixed-use and multiple-use development is an increasingly high-profile option for land use planning. For this reason, adequate assessment of the capacity yielded by such development, including the expected mix of types and relative amounts of space devoted to each use within projects, is likely to gain as a priority within LSCM.

- The capacity of industrial and commercial lands can be assessed from building utilization and land consumption rates. However, the methodology for doing so is problematic. Firm sizes vary considerably even within commercial and industrial sectors, and there is a less predictable relationship between employment levels per sector and land utilization than between households per housing type and land needed to accommodate them. Over time economic growth will result in different patterns of land consumption for different types of firms. Finally, employment may increase within single firms without any increase in land consumption at all, as the new employees are incorporated within existing facilities or sites. (See Chapter 5 for a more complete discussion of these issues.)

- Capacity analysis should accurately account for the "under-build" phenomenon (see Figure 2.3), especially in single-family residential develop-

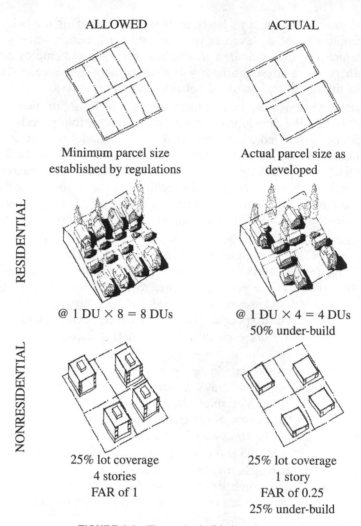

ALLOWED   ACTUAL

Minimum parcel size
established by regulations

Actual parcel size as
developed

RESIDENTIAL

@ 1 DU × 8 = 8 DUs

@ 1 DU × 4 = 4 DUs
50% under-build

NONRESIDENTIAL

25% lot coverage
4 stories
FAR of 1

25% lot coverage
1 story
FAR of 0.25
25% under-build

FIGURE 2.3   The under-build phenomenon

ment. It is well established that vacant land is often developed at densities lower than allowed by regulation—this can occur under a variety of regulations and even in urban areas (Real Estate Research Corporation 1982; Knaap and Nelson 1992). In cases of residential under-build, developers' profits are larger for large lot and large house products, and homeowners have apparently been willing and able to pay higher prices than anticipated by land regulators. Furthermore, such areas are unlikely to undergo infill or redevelopment within the usual planning time horizons because the subdivisions are new and the value of improvements relatively high. In commercial development, accounting for under-build is also an issue, largely stemming from a historical oversupply of commercially zoned land in many jurisdictions. Studies of nonresidential land supply and capacity, including commercial land uses, recently undertaken by local and regional governments using GIS (see Chapter 4 and Montgomery County, Maryland, in Appendix B), have begun to grapple with the challenges associated with predicting future commercial build-out levels.

- Longitudinal data on development outcomes under various zoning and other land use regulations, as well as for various land uses (e.g., housing types and industrial sectors), would greatly improve analyses by providing an empirical basis for capacity assumptions. Some jurisdictions have employed trend data (such as observed development type and density within zoning categories, subareas, or land use subtypes) to appropriately discount expected densities to account for under-build. For example, the City of Fitchburg, Massachusetts, utilized land information collected over 15 years to analyze single-family residential under-build. This study identified roadways, physical constraints (such as steep slopes and wetlands), and inefficient platting as major determinants of the phenomenon, and provided information used to establish an under-build factor (58 percent of maximum allowed dwelling units per acre) for future land capacity work. In a broader effort, the State of Washington's Buildable Lands Program directs a multicounty effort to collect development data annually to guide periodic analyses and assessment of land supply and capacity (see Appendix B).

- Minimum-density zoning has been proposed as an innovative way to address under-build and has thus far been applied in a handful of jurisdictions throughout the country, notably Renton, Washington, Portland, Oregon, and Sacramento County, California (Morris 1996). To the extent that such zoning is applied more broadly in the future, minimum densities offer an alternative standard against which to measure the anticipated capacity of a buildable land supply.

- Improved methods are needed for doing small-area buildable land supply and capacity studies, particularly approaches adequate to assess local demand and anticipated development levels in relation to regional growth forecasts. However, the uneven availability of disaggregated data on both supply and demand, makes the analysis of land capacity problematic, especially for small suburban jurisdictions, or for neighborhoods and small planning areas within larger jurisdictions.

- Market factor adjustments to capacity effectively reduce expected numbers of dwelling units or building square footage totals to be yielded by private development of the land supply. They usually represent percentage deductions from capacity figures for part or all of the land supply. However, neither any of the specific factors employed, nor the theoretical justification for applying a factor at all, has been thoroughly supported by empirical evidence. Given good data on past and current development practices and an appropriate methodology for forecasting future demand and development patterns and densities, an area-wide market factor may not be appropriate. It may, in fact, be more important to account for local variability in market conditions, policies, and social values that have an impact on land development (Enger 1992). This would include considering the effects of neighborhood opposition to development or redevelopment of land to densities and uses that, although allowed by zoning, are considered disruptive to neighborhood "scale" or "character." For these reasons, subarea-specific market factors may be preferable to areawide adjustments.

As an overview of Elements 2 and 3, Figure 2.4 summarizes the entire analysis of buildable land supply and development capacity. From left, the figure illustrates the breakdown of the total land base into constituent supply components: fully developed parcels, committed lands in the "pipeline," and the three types of buildable land. Buildable lands are converted to development capacity amounts, which are then aggregated to tally total net capacity. Meanwhile, constraints effectively operate as deductions from supply (for prohibitive factors), capacity (for mitigating factors), or total net capacity (for areawide or market-factors). The discussion now considers contextual issues in LSCM.

Element 4: Information for Plan Making and Plan Implementation

The first three elements are the basic methodological components of LSCM. The fourth element locates LSCM within a system for comprehensive land use planning (see Figure 2.5) that encompasses data collection, forecasts and analyses of the supply of and demand for land, and provision of information to guide land use policies and regulations. Taking the model as a whole, two overarching dimensions deserve emphasis. First, measures of land supply and capacity derive their significance in relation to measures of demand. Supply and demand analyses are ideally carried out interactively, effectively modeling relationships between them. Second is the dimension of time. Fully implemented, land supply monitoring represents a continuous iterative process of measurement, analysis, evaluation, and public action. Figure 2.5 represents a single loop that may be repeated comprehensively every three to five years (or more), and partially on a more frequent basis. This section focuses on these two dimensions.

In several important respects, land monitoring is given shape and purpose by a set of plans and related planning activities. First, land use plans give a normative cast to predicted patterns of regional demand for land and improvements. They do so by guiding the allocation of regional population and employment forecasts to small subregional areas (see Figure 2.6). Second, plans are essential in defining both the time frames and the specific areas over which supply and capacity are evaluated relative to demand. Although plans cannot

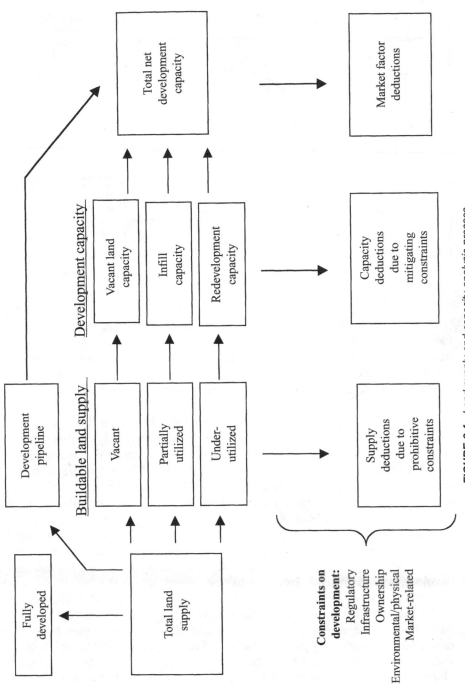

FIGURE 2.4 Land supply and capacity analysis process

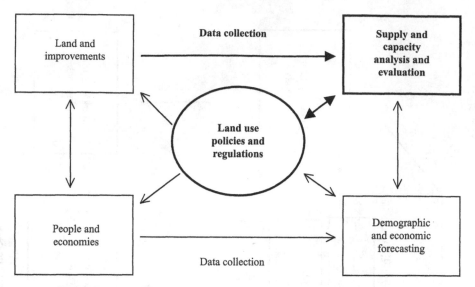

FIGURE 2.5 Land monitoring within a system for comprehensive planning

be considered to define demand per se, they embody translations of demand figures into desired land uses and land use distributions. Notably, opponents of comprehensive land use planning often focus their criticisms on the perception (or fact) that plans do not respond to actual demand. By extension, critics of supply and capacity analyses often begin by questioning the very plans by which these activities are governed.

In the planning process, a judicious allocation of regional demand within a given area and over a specified future time horizon will rely on two aspects of land monitoring: an account of how land supply may constrain the capability to accommodate that demand, and an assessment of how specific land use objectives are achieved subsequent to that allocation. Insofar as the supply of land and improvements is perceived as "limited," land monitoring thus plays a crucial role in both plan making and implementation processes. As such, land supply monitoring becomes part of a comprehensive planning support system—a key tool for generating planning "intelligence," which may be used to model policy alternatives and their effect on future land use and development, and provide feedback to government about the effects of plans as implemented over time (Webber 1965; Klosterman 1997; Heikkila 1998).

As the term *monitoring* implies, the evolution of LSCM as an integral element of urban and regional planning and plan evaluation may have the most profound impact by providing a flow of data and analysis outputs flexibly and within a continuous time frame. As emphasized by Godschalk et al. (1986), planners have tended to view land supply as a static spatial concept rather than a dynamic socio-economic one. In contrast, LSCM, especially at the parcel level, allows planners to monitor the dynamics between the demand

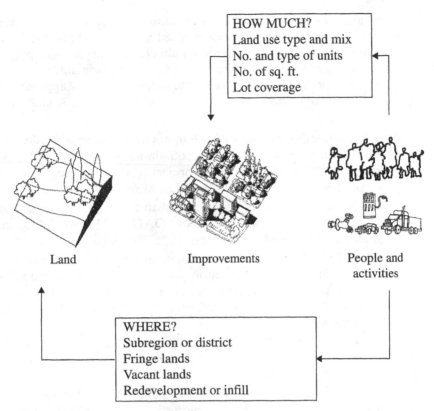

FIGURE 2.6   *The demand allocation process—Implications for supply and demand*

for land and improvements and the supply of land and improvements as influenced by regulatory frameworks governing land utilization. Chapter 3 describes one such system in action and argues for establishing an ongoing land supply inventory management system as a tool for calibrating regulatory and fiscal policy in line with a balance between land supply and demand.

Following on and supporting these arguments, we define land monitoring as implying a dynamic relationship between plan making and plan implementation. First, plan making entails establishing baseline data for a comparative evaluation of multiple courses of action—policies, regulations, and public investments—that focuses on both intended and unintended impacts. As a related exercise, the testing of various scenarios involves assessing the performance of proposed plans and regulations under a range of potential future conditions. Second, plan implementation addresses the translation of policy plan goals and objectives into specific regulations and public actions that affect desired outcomes (for example, alternative urban forms, reduced sprawl, higher densities, mitigated environmental impacts, affordable housing, lowered infrastructure costs). LSCM can provide information necessary for land management and private-sector activities. Monitoring may serve to answer questions such as, Is the plan being implemented as designed? Has the plan, as implemented, had the desired or expected effect? Are targets and specific objectives being met? Where, and where not? How, and how not?

Our survey of planning agencies revealed the following types of applications of LSCM to plan making and implementation (see Appendixes A and B):

- Delineation and administration of urban containment areas, either by demarcating or adjusting urban growth boundaries (as by Portland Metro and Washington State jurisdictions) or projecting urban service areas (as by Metropolitan Council of the Twin Cities, Minnesota)
- Guidance in the allocation of regional growth forecasts (or targets) to subregional areas—as, for example, by SANDAG, San Diego, California, through an allocative modeling process
- Monitoring of UGB performance in accommodating population and employment growth over time—emphasis given to achieving desired locations and densities of developments without unduly constricting supply and causing land and housing price inflation—as in the Washington State Buildable Lands Program
- Assessment of the impact of local land use regulations (e.g., zoning or more innovative growth controls, such as adequate public facilities ordinances) on supply and capacity
- Assessment of potential amounts, patterns, and locations of future development within subregional areas of interest (e.g., watersheds, planning areas, and transportation nodes and corridors)
- Specialized applications, such as monitoring and screening land for transfer of development rights (TDR) programs, conducting inventories of vacant or publicly owned (often tax-delinquent) lands for economic development purposes, and natural hazards risk management

Extending across the range of applications of LSCM, the following are major issues to be considered as they relate to plan making and plan implementation:

- A valid baseline must be established for existing land use, land supply, and capacity from which to assess different scenarios for the future and as a basis for monitoring changes in land utilization over time.
- LSCM can be enhanced through coordination with short- to medium-term planning, such as medium-term capital improvements programming. However, over the medium term, difficulties arise in assessing capacity with respect to differential growth characteristics that may exist between subareas that lie near or adjacent to jurisdictional boundaries. Applications of capacity analysis to subarea transportation planning are particularly prone to such problems.

- Established routines for ongoing monitoring of land supply and capacity extend LSCM beyond the sporadic efforts that correspond to planning cycles and the episodic data requirements of one-time studies. Portland Metro's "Performance Measures," King County's Benchmarks Program, and Washington State's Buildable Lands Program (see Appendixes A and B) are examples of the formalization of processes for continuous monitoring and analysis in support of local and regional planning.

- Jurisdictions should seek concordance between measurable plan objectives and the level of resolution of land monitoring—essentially addressing the complexity of translating elements of long-term strategic planning into the units and scales that fit with existing practices of monitoring and analyzing land supply data (and vice versa). For example, the Buildable Lands Program in Washington State will have to address land supply and capacity within counties, cities, urban growth areas, and designated urban centers. To the extent that the parcel is adopted as a unit of collection and analysis in all of the participating jurisdictions, it will be possible to achieve flexible aggregation of land supply and capacity to these various subareas of interest.

- Monitoring "pipeline development" by tracking new projects through the approval and development process provides data for projecting future land consumption patterns and densities over the near term. Such information can also be used for estimating short-term supply and demand balances and to anticipate the response of the private development industry to spikes in demand, especially for housing. Other applications of pipeline monitoring include tracking development activity in relation to medium-term infrastructure and service provision, administering adequate public facilities ordinances, and tracking the build-out of subdivisions and large multi-phase projects.

- Land supply inventories are important not only to ensure that there is enough land, but also to identify situations where there is an excessive supply of land for a given use. LSCM may thus be used to provide justification for targeted rezones or other corrective actions to address oversupply problems.

- Establishing criteria for a geographic allocation of growth entails translating population and employment projections or forecasts into future patterns of land use (see Figure 2.6). Relatively accurate population and employment forecasts can be made for large geographic areas, but their allocation to individual areas within a region can be problematic. Straight-line projections applied to subarea growth allocation may not be valid, given the influence of changing local markets, federal regulations, and other factors. Furthermore, there is a tension between regional planning, which commonly follows top-down allocation of demand from regional population and employment projections, and the practices of land management and private land development, which respond primarily to local supply and demand factors and operate at the regional subarea, district, neighborhood, or even site level.

- Translating population and employment growth into demand for specific types and configurations of development is also often done in a straight-line fashion—such as by proportioning residential market shares or projecting household size or square-feet-per-employee estimates into specific space requirements. Information derived from land monitoring can be used to refine these assumptions and may entail tracking demographic, economic, and development trends, as well as assessing the degree to which the market is supporting, and the development industry producing, new "types" of land development (for example, clustered multifamily housing, mixed-use projects, or high-tech industrial parks).

- Computerized urban simulation models can add sophistication to subregional demand allocation by accounting more explicitly for dynamic factors that affect land use change such as land prices and the behavior of the development industry, the preferences of households and businesses as consumers of development products, and interactions with transportation investments. Chapter 8 describes one such model and its potential to be used as an enhancement to LSCM.

- LSCM should address factors that affect the comparability of repeat measures of supply or capacity over time, which include the following:
  - Changes in methodology and basic analysis assumptions
  - Changes related to data and analysis tools used
  - Policy shifts (e.g., zoning, and infrastructure investment)
  - Changing patterns of land consumption by new development (e.g., real estate and construction cycles)
  - Changing political boundaries (e.g., annexations, incorporations)
  - Variability in demand responding to economic cycles

- Highly prescriptive mandates from state and regional governments for how land monitoring and analysis should be performed result, in many cases, in attempts to impose, from the top down, specific methods and database designs. On one hand, this may conflict with local situations and available planning resources. On the other hand, there is a clear need for some form of standardization among jurisdictions in developing appropriate methods for monitoring and analyzing land supply and capacity (see Chapter 7).

- To the extent that LSCM is carried out on a regional level, it is important that procedures be developed to integrate knowledge from local planning departments and other agencies. This is relevant on several levels. First, local knowledge will provide multiple sources for correcting errors, inconsistencies, or missing data in regional land supply databases. Second, analysis assumptions may be partly based on input from local jurisdictions, where knowledge of subregional markets and land availability can produce refined and location-specific estimates of supply and capacity.

SANDAG and the Cape Cod Commission are two cases in which local input has played a significant role in regional database development and analysis efforts (see Appendix B).

CONCLUSIONS

The four elements of land supply and capacity monitoring described in this chapter provide a conceptual framework to address the various dimensions of LSCM practice. They draw from the approaches of some of the most sophisticated land monitoring programs developed to date. The relevant issues discussed for each element reflect primarily the experiences of local and regional governments in implementing these programs. They also include the general methodological challenges outlined in Chapter 1, encompassing the need to establish comprehensive databases, common definitions, credible methodologies, and general agreements among a broad array of actors and interests in the public and private sectors.

Advances in land information technology are predicted to continue, and it is likely that it will soon be possible to readily capture and analyze detailed longitudinal data on land supply and capacity. Improvements in the ability to describe and interpret trends (both short- and long-term) in the use of urban land will necessarily provide better information on ongoing and ever-changing conditions of land development. Such information, in turn, can help refine land use and infrastructure plans and calibrate them to local and regional conditions.

Over the coming years an increasing number of local governments are likely to undertake land supply monitoring, and an increasingly broad range of planning and other local government functions will incorporate techniques of land supply and capacity analysis. To ensure the success of their efforts, local officials should strive for clarity about the scope of these monitoring programs, the definitions of major elements in the process, and the range of valid interpretations that may be derived from the results—both quantitative and spatial—of this work.

REFERENCES

American Planning Association. 1999. LBCS Manual. Available May 10, 1999 at [http://planning.org].

Anderson, Larz T. 1987. *Seven methods for calculating land capability/suitability*. Planning Advisory Service Report Number 402. Chicago: American Planning Association.

Bollens, Scott A. 1998. Land supply monitoring systems. In *The Growing Smart working papers*. Vol. 2. Chicago: American Planning Association.

Dale, Peter F., and John D. McLaughlin. 1988. *Land information management: An introduction with special reference to cadastral problems in third world countries.* Oxford: Clarendon Press.

Dearborn, Keith W., and Ann M. Gyri. 1993. Planner's panacea or Pandora's box: A realistic assessment of the role of urban growth areas in achieving growth management goals. *University of Puget Sound Law Review* 3: 975–1024.

Dueker, Kenneth. 1998. Telephone interview by author. January 8.

Easley, V. Gail. 1992. *Staying inside the lines: Urban growth boundaries.* Planning Advisory Service Report No. 440. Chicago: American Planning Association.

Enger, Susan C. 1992. *Issues in designating urban growth areas.* Parts I and II. Olympia, Wash.: State of Washington, Department of Community Development, Growth Management Division.

Godschalk, David R., Scott A. Bollens, John S. Hekman, and Mike E. Miles. 1986. *Land supply monitoring: A guide for improving public and private urban development decisions.* Boston: Oelgeschlager, Gunn & Hain, in association with the Lincoln Institute of Land Policy.

Harris, Christopher. 1992. Bringing land use ratios into the '90s. *PAS Memo.* 8: 1–4.

Heikkila, Eric J. 1998. GIS is dead: Long live GIS! *Journal of the American Planning Association* 3: 350–360.

Kaiser, Edward J., David R. Godschalk, and F. Stuart Chapin. 1995. *Urban land use planning.* Urbana and Chicago: University of Illinois Press.

Kivell, Philip. 1993. *Land and the city: Patterns and processes of urban change.* London and New York: Routledge.

Klosterman, Richard. 1997. Planning support systems: A new perspective on computer-aided planning. *Journal of Planning Education and Research* 17: 45–54.

Knaap, Gerrit, and Arthur C. Nelson. 1992. *The regulated landscape: Lessons on state land use planning from Oregon.* Cambridge, Mass.: Lincoln Institute of Land Policy.

Korte, George. 1997. *The GIS book.* Santa Fe: OnWord Press.

Moore, Terry. 1997. Telephone interview by author. October 12.

Morris, Marya. 1996. *Creating transit-supportive land-use regulations.* PAS Report No. 468. Chicago: American Planning Association.

Public Technology, Inc., Urban Consortium, and International City Management Association. 1991. *Local government guide to geographic information systems: Planning and implementation.* Washington, D.C.: Public Technology, Inc., Urban Consortium, and International City Management Association.

Real Estate Research Corporation. 1982. *Infill development strategies.* Washington, D.C.: Urban Land Institute and American Planning Association.

Webber, Melvin M. 1965. The roles of intelligence systems in urban-systems planning. *Journal of the American Institute of Planners.* 11: 289–296.

Part II

Case Studies

3

Portland, Oregon: An Inventory Approach and Its Implications for Database Design

Lewis D. Hopkins and Gerrit J. Knaap

Local governments have always played a major role in the urban development process. They provide infrastructure, issue subdivision and building permits, plan and zone land use, and assess and collect taxes. To perform these tasks, local governments need extensive and current information on land use, property ownership, and urban development events. In the past these needs have been met via disparate and disjointed information systems, including assessors' taxation files, parcel plat maps, building and subdivision permit records, and various planning and zoning maps. Such disjointed information systems are inadequate, however, to support comprehensive land use planning and growth management.

Geographic information system (GIS) technologies have enabled the integration of spatially referenced data, such as zoning and tax-lot maps, with tabular data such as assessors' files and building permit records. As a result, many communities now have access to detailed data and, in particular, data on tax-lot parcels. However, the easy availability of parcel data may create

Monitoring Land Supply with Geographic Information Systems, edited by Anne Vernez Moudon and Michael Hubner ISBN 0 471371673 © 2000 John Wiley & Sons, Inc.

unrealistic expectations of what can and cannot be done in using the attributes of parcels for land supply monitoring. The Portland, Oregon, metropolitan area provides an example of land monitoring and GIS database development. Stimulated by a state mandate to manage metropolitan urban growth, Metro, the regional land use authority, developed a regional GIS that contains attributes of tax-lot parcels as well as attributes of other spatial units. Called the Regional Land Information System (RLIS), it represents perhaps the most advanced metropolitan region GIS in the United States and is held up as a model for further development of GIS applications in land supply monitoring (see Chapter 1).

In this chapter we discuss the use of comprehensive land information systems for land use planning and growth management based, in part, on our experience in using RLIS. We begin by describing RLIS and how it has been used for managing urban growth in metropolitan Portland. Next we discuss conceptual issues in the use of land information systems for growth management, including land supply monitoring and land use planning. Growth management is framed as a problem of inventory control, and implications for database design are considered through a comparison of two ways of structuring data in land information systems. The comparison yields strong suggestions as to when and where parcel data are most useful in the land monitoring process. The chapter concludes with an assessment of the use of RLIS in managing urban growth and recommendations for further development of such systems.

THE REGIONAL LAND INFORMATION SYSTEM AND ITS USE FOR MANAGING PORTLAND METROPOLITAN GROWTH

Land use and urban growth management are long-standing issues of concern in the Portland metropolitan area. In 1973, the State of Oregon required all cities and counties to prepare comprehensive land use plans and to designate Urban Growth Boundaries (UGBs) around all urban areas. By state law, UGBs must contain sufficient land for 20 years of urban development and must be reviewed every 5 to 7 years. Responsibility for maintaining the Portland area UGB belongs to Metro, the only directly elected regional government in the United States (see Chapter 6 and Appendixes A and B).

Metro first established the UGB in 1979, which changed little over the subsequent decade of slow urban growth. As growth accelerated in the late 1980s and early 1990s, Metro began an extensive reevaluation of the UGB and launched an ambitious planning process designed to shape the direction of urban growth for the next 50 years. In part to facilitate these tasks, Metro created RLIS.

The Regional Land Information System

RLIS is a comprehensive geographic information system that contains more than 100 data items. The data include tax lots; floodplains, slopes, rivers, and other natural features; population, income, and other census data; county, city, school district, and other political jurisdictions; zoning, plan designations, urban growth boundaries, and other regulatory constraints. Most of the data cover all of Washington, Multnomah, and Clackamas Counties. They are updated at various intervals, often through cooperative arrangements with local governments (Table 3.1).

Two data items deserve special mention. First, the tax-lot parcel geography data, originally developed by Portland General Electric, are maintained by the three county governments, which collect subdivision data and maps from local governments to keep the data current. Parcels are identified by the tax-lot numbers used by the county assessors' offices. Second, the vacant land data are maintained by Metro. This GIS coverage was developed with the use of orthophotography and is updated annually. Parcel boundaries are overlaid on aerial photographs to identify parcels as vacant, under construction, or developed. Portions of parcels greater than one-half acre in size can also be classified as vacant. Geocoded building permits are used to flag areas of development activity and focus photo interpretation on areas of likely change.

The Buildable Lands Inventory and Capacity Analysis

Metro conducted a Buildable Lands Inventory and Capacity Analysis in 1997 to comply with state statutes and to further the regional planning process. The process began with the vacant land coverage, which provided an estimate of most of the land available for development. From this estimate, subtractions were made for land currently under development and for land difficult to develop because of environmental constraints. The development capacity of the remaining land was estimated by computing the maximum development density allowed by zoning regulations (this process is described in more detail in Appendix B). Based on Metro's land inventory and capacity analysis, the Metro Council recommended only minor adjustments to the UGB, despite concerns expressed by home builders and other interest groups about land and housing price inflation.

The design and use of RLIS for managing growth involves many conceptual and technical issues. This chapter, however, focuses on two: How should land monitoring systems be used to manage growth? How should data be organized in land monitoring systems? To address these issues, conceptual frameworks are developed for monitoring land supply and for structuring data in land information systems; within these frameworks, the use and design of RLIS are examined.

TABLE 3.1 The Regional Land Information System

Category	Coverage(s)	Source(s)	Dates/Updates	Spatial Unit(s)
BOUNDARY	City limits, counties, Metro, neighborhood organizations, school districts, Tri-Met, Urban Growth Boundary, voter precincts, zip codes	City and county plans and ordinances, Multnomah County Assessor, Portland General Electric, Metro, U.S. Postal Service, county elections office	Some updated quarterly; others are always current; some have not been updated for more than a year.	Political jurisdictions
CENSUS	Census tracts and block groups, employers	U.S. Census Bureau	1990 Census data; employers dated 1994	Census units
ENVIRONMENT	Contours, environmentally significant zones (Portland only), soils, vegetation cover	Metro, U.S. Geological Survey, Soil Conservation Service, Pacific Meridian Resources	Dated 1991 to 1996	Environmental areas
LAND	Building permits, generalized regional plan designations, generalized regional zone designations, local plan designations, local zoning designations, parks and open space	Building permits issued by counties, Metro, city and county comprehensive plans and zoning ordinances	Parks date to 1995; others are all updated quarterly.	Point location, planning areas, polygons

PLACES	City halls, fire stations, hospitals, schools	Metro Regional Directory, Thomas Guide, Portland Board of Education, Tualatin Valley Fire and Rescue, Clackamas County Green Map, Washington County, local jurisdictions	Schools date to 1996; all others to 1994.	Point location
STREETS	All streets, all arterial streets, major arterial streets	Oregon Department of Transportation (ODOT), U.S. Geological Survey (USGS), Metro, U.S. Census Bureau, local jurisdiction map books	All streets updated quarterly; others dated 1995–1996.	Line location
TAX LOTS	Quarter sections, sections, tax lots, townships	Cities of Beaverton, Milwaukee, Oregon City, and Tigard; Multnomah County Assessor, Portland General Electric; Metro	Tax lots updated quarterly; all others dated 1992.	Tax lots
TRANSIT	Bus system, light rail, Park & Ride lots, railroads	TriMet, Metro	Dated between 1992 and 1997	Line location
WATER	Floodplains, major rivers, rivers, streams, watersheds, National Wetland Inventory	Metro, ODOT, USGS, U.S. Army Corps of Engineers, U.S. Fish and Wildlife Service	Dates vary across wide range (1974–1996).	Environmental areas

CONCEPTUAL ISSUES IN MONITORING LAND SUPPLY

Although the process of urban growth and development involves actions by numerous actors at particular times and places, it can be viewed as a process in which the "state" of land changes through a sequence of several discrete events. Such a process is illustrated in Figure 3.1. In the illustration, a farm is sold to a developer and farming stops. The developer then subdivides the land into residential lots and obtains sewer and water services. Subsequently, homes are built and residents move in. Often, though not always, changes in the state of land are coincident with a change in ownership; for example, ownership passes from a farmer, to a developer, to a builder, to a home owner. The length of time land remains in various states may vary, as may the sequence of states. Much, if not most, of the land development occurs, however, through the process characterized in Figure 3.1.

Land Inventories and Monitoring Systems

Land inventories and development monitoring have always been critical elements of the land use planning process. Segoe (1941, 54), for example, states:

> The preparation of a city plan requires . . . the assembly of information and data about land utilization, population, public facilities and services of all sorts,

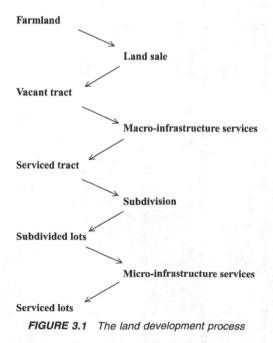

Farmland

Land sale

Vacant tract

Macro-infrastructure services

Serviced tract

Subdivision

Subdivided lots

Micro-infrastructure services

Serviced lots

FIGURE 3.1   The land development process

street traffic, railroad traffic, and others, most of which have to be portrayed on maps for analysis and appraisal of conditions and trends.

Such simple inventories represent spatial snapshots of the state of land in the local jurisdiction. When the plan is revised, another snapshot must be taken. Thus, the traditional planning process involves a sequence of snapshots of the state of land, which are conducted in cycles that correspond with cycles in the planning process. From such a sequence of snapshots, it is possible to infer the development events that took place between snapshots. The events themselves, however, are not of primary interest and are not explicitly recorded for planning purposes.

The term *land monitoring system* became popular in the mid-1980s and is now used to describe systems that include both periodic land use inventories and records of land development events. According to Godschalk et al. (1986), such systems, which they call Automated Land Supply Information Systems (ALSIS), contain two types of files:

1. A parcel file, derived from the assessor's records and containing information on land ownership, value, and use for each parcel in the jurisdiction
2. Project files, derived from applications for subdivision plan, planned unit development, and site plan approval, and containing data on each proposed project and its stage in the process of development review

Furthermore, the authors suggest that separate files may be kept for other purposes, such as tracking building permits, sewer permits, and environmental conditions. Systems of the type described by Godschalk et al. incorporate data on the "state" of land in the parcel file and data on development "events" in the project files, although the two are not systematically integrated.

The state of the art in land development monitoring advanced substantially in the 1990s with the rapid adoption of GIS technology. With GIS technology, it becomes possible to integrate data in an ALSIS with computerized maps. That is, data on each parcel can be stored as a polygon in space, with attributes that include state characteristics and a history of development events. Such parcel event histories can also be used to animate images to illustrate the process of urban development (Ding et al. 1997).

Inventory Management

Although advances in land information systems have occurred rapidly and at an accelerating pace, conceptual advances in how to use such systems for land use planning and growth management have been slow. Information systems should be designed to serve the purpose for which they are needed. As discussed by Godschalk et al. (1986), the primary purpose of developing a land information system is to facilitate land and infrastructure management:

Land supply information is the "missing link" in many critical development decisions. Public policies regulating the amount of land available for development made without the benefit of accurate land inventory knowledge can have disastrous effects on the price of raw land or the efficiency of providing public facilities. Private development decisions made without timely knowledge about land supply in the local market can drive up housing costs unnecessarily. Such imperfect information multiplies the risk and uncertainty of both public and private development decisions (Godschalk et al. 1986, 2).

Thus, according to these authors, the principal purpose of developing land information systems is to monitor and maintain an adequate inventory of land and to ensure the timely provision of infrastructure. Although virtually absent in the growth management literature, problems in inventory management have been addressed in the literature on business administration and operations research for quite some time (see Sippen and Bulfin (1997) for a brief literature review).[1] The following sections present some fundamental concepts in inventory management, from which an analysis of database design is begun.

The Perpetual Inventory Method

The perpetual inventory method is the most common method of obtaining and storing information for inventory management (Greene 1969). This method involves a continuous accounting of incoming materials, outgoing materials, and the balance of materials on hand. It begins with an explicit list and definition of items to be managed and an initial measurement of the stock on hand. For a land information system, this approach will require a careful definition and an initial measurement of the acres of land in various states, then a perpetual accounting of events whereby land moves from one state to another.

The basic concept of the perpetual inventory is illustrated in Tables 3.2a and 3.2b, which show transaction records for two inventories of land zoned for single-family residential use: (1) an inventory of such land that is within the Urban Growth Boundary but is not subdivided and (2) an inventory of subdivided single-family residential land. Land is added to the first inventory when area is added to the UGB, and is subtracted when land is subdivided. Land is added to the second inventory when it is subdivided, and subtracted when building permits are approved. Critical to the effective use of the perpetual inventory method is clear and logical specification of the inventory organization, including a carefully defined classification system, unique parcel numbers (or other unique identifiers for other units of stock), and verification processes. A significant advantage of the perpetual inventory method is "continual verification." The record of changes in inventories continually verifies

---

[1] Similar problems are reported in the literature on capital improvements programming (Haynes et al. 1984) and in the literature on planning theory (Intriligator and Sheshinski 1986; Knaap et al. 1998).

TABLE 3.2a   Inventory Record for Residential Land Within Urban Growth Boundary

Date	Event	Received	Shipped	Balance
	Initial balance			1,000 acres
10/2/98	Add residential to UGB	100 acres		1,100 acres
12/10/98	Subdivision approval		60 acres	1,040 acres

the status of each inventory. As shown in Tables 3.2a and 3.2b and Figure 3.1, land moves through a sequence of inventories, as a sequence of events changes land from one state to another. Each of these inventories may be described by multiple attributes. For example, the land at one state may be vacant, zoned for residential use, subdivided into lots, and serviced. The stock of such land would be changed by any event that changed any one of these attributes.

The Economic Order Model

Incorporating a perpetual inventory approach as part of a land information system has several advantages, which include the systematic integration of information on development events with information on development states, and the continuous verification and adjustment of the inventory. Perhaps the most significant advantage, however, is the opportunity to use well-established, formal methods of inventory control. The simplest method of inventory control is the economic order model (Sippen and Bulfin 1997).

The economic order model is illustrated in Figure 3.2. Time is depicted on the horizontal axis, and the inventory—or the quantity in stock—is depicted on the vertical axis. The sawtooth pattern in Figure 3.2 reflects a "reorder policy" in which the inventory is replenished instantaneously when the stock falls to zero and by the same quantity at each time. This example assumes that the depletion rate of the inventory is constant. In this simple model, the average inventory equals one half the order quantity. The optimal order size depends on the cost of replenishing the inventory, the cost of carrying the inventory, and the depletion rate.

The economic order model in Figure 3.2 depicts the simplest model of inventory control. More complex models for managing urban land and infrastructure include factors that capture uncertainty in the depletion rate, potentially long lead times needed to replenish the stock, and rapidly increasing

TABLE 3.2b   Inventory Record of Subdivided Residential Land

Date	Event	Received	Shipped	Balance
	Initial Balance			200 acres
12/10/98	Subdivision approval	60 acres		260 acres
1/25/99	Building permits		15 acres	245 acres

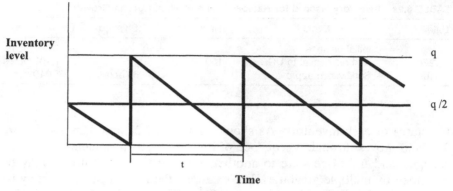

FIGURE 3.2   *An inventory reorder pattern*

costs when the stock falls below a critical level. These complexities are not discussed here (see Sippen and Bulfin (1997) for a complete discussion). Instead, two points are emphasized: (1) Sound land and infrastructure management requires that information be organized so as to support inventory control models; and (2) the failure to organize information accordingly has led to confusion regarding issues such as the frequency with which land use regulations should be revised and the appropriate size of urban growth boundaries.[2]

Markets and Market Models

Unlike many inventory management problems, which are internal to a firm, the problem of managing land and infrastructure involves market responses to inventory policy. For this reason, it is important to maintain information on prices as well as on stocks. As in the conduct of monetary policy, it may be more effective to make supply decisions by monitoring price than by monitoring supply. That is, it may be more efficient to increase the stock of vacant land within the UGB when the price of land, not the stock of land, reaches some critical level.[3]

---

[2] Issues concerning the appropriate size of a UGB are often addressed in terms of a "market factor," which is intended to provide precautionary excess supply and is often expressed as a percentage of the estimated demand for land over the entire planning horizon. In the economic order model such precautionary supply is expressed in terms of "days of inventory." One interpretation of the Oregon legislative mandate that the UGB be revised every five years to include a 20-year supply of land, is that 15 years is the prescribed precautionary supply, in which case increasing this further by a market factor is inappropriate.

[3] To manage the money supply, the Federal Reserve Board once closely monitored money supply data and made monetary policy decisions accordingly. In recent years, the Federal Reserve Board closely monitors interest rates (the price of money) as the basis for monetary policy.

DATA STRUCTURES TO SUPPORT AN INVENTORY APPROACH

As described in the previous section, land monitoring for growth management can be treated as a standard inventory problem, providing information to decision makers about when, where, and by how much the stock of buildable land should be "replenished" to meet expected demand. To apply such an approach to growth management, however, requires that land information systems be structured to address inventory problems.

Providing built forms for human settlement involves inventories of several different "products" (different land uses at different densities) and inventories of different stages in the development process. Residential uses, for example, come in different densities, income levels, market segments, and geographic locations. Commercial and industrial uses are also differentiated. Each stage in the sequence of development from resource land (agriculture, forestry, or extensive recreation), to speculatively held land, to macroserviced land (arterial roads, interceptor sewers), to subdivided land, to microserviced land (access roads, local sewers, schools, parks), to built form, to occupied unit, is an inventory.

If the transitions from one inventory to the next are defined, then an inventory model can be used to analyze land in any one inventory or in any set of inventories. This decision will depend in part on ease of recording the information on transition events, and in part on the usefulness of lead times in making reorder decisions. It may be desirable to know what land is currently under option by developers so as to identify earlier in the sequence land that is likely to be developed. It would be very difficult, however, to collect such data, so information on subdivision applications will probably suffice.

Database Requirements

The crucial point of the proposed approach is to use an inventory management system that focuses on *changes* in stock rather than on collecting snapshot data about the stock of land as a one-time or recurring task. The focus on transition events affects the design of databases and information systems that support such land supply monitoring. The inventory model requires a database that allows us (1) to determine the status of an inventory at any arbitrary time, (2) to determine rates of change in status, and (3) to pass attributes of stock from one stage of development to the next.

The standard approach to database design identifies entities, relationships among these entities, and attributes of these entities. An entity is any object (tangible or intangible) that is pertinent to the situation at hand. The most crucial entities for this purpose are the spatial units of land that are accounted for in the inventory model. These may be tax-lot parcels, infrastructure service areas, the UGB, areas zoned for particular uses, or any of a number of other possibilities. These entities account for the status of stock, changes in stock, and attributes of the stock.

If the spatial entity used as the unit of accounting for the stock of land changes through time, then the database must be able to translate from the set of land areas before the event to the set of land areas after the event. If we order in boxes of parts and ship out assembled products, we need to know how many of each part are in each assembled product in order to translate from a count of product shipped to a count of reduction in parts on hand. If we stock large parcels of land and develop (i.e., ship) subdivided parcels, we need an analogous translation.

Such changing entities also create difficulties in passing attributes from one entity to the next. There is no direct way to distribute the attributes of a parent unit of accounting to descendant units of accounting unless the parent attributes are homogeneous geographically. Even if a data structure tracking parents and descendants were implemented, geographically heterogeneous attributes passed through the transition may be invalid without entry of new data, based on human interpretation, to assign attribute values to the descendant units.

Entities should therefore be defined so that all attributes apply homogeneously to the entire entity. For example, the attribute value of "built" or "vacant" may not apply homogeneously to an entire tax-lot parcel, because part of it may be developed and part not developed. On the other hand, "owner name" is an appropriate attribute of a tax-lot parcel, because the owner (or owners) owns the entire parcel. The principle is that an entity should be homogeneous with respect to any attribute of that entity. Spatial entities (points, lines, polygons, or grid cells) are frequently encoded as areas homogeneous with respect to a particular attribute. Tax-lot parcels are, by definition, areas that are homogeneous with respect to owner. Other attributes may share the same geographic area, so that a particular entity is homogeneous for several attributes. Parcels are appropriate not only for owner, but also for such attributes as assessed value of land, parcel creation date, and sale date. To use attributes that do not apply homogeneously to the same entity, however, we should derive a Homogeneous Geographic Area (HGA) for the combination of attributes. A derived HGA is simply the largest geographic area that results from overlaying (or logically intersecting) the HGAs of each individual attribute. To derive an HGA from coverages of regulatory zones, infrastructure service areas of various kinds, and environmental constraints of various kinds, we can use GIS to identify the geographic areas that are homogeneous with respect to combinations of these attributes.[4]

---

[4] This explanation of derived HGAs is equivalent to the "factor combination" approach as described in Hopkins (1977). Others have referred to these as "terrain units" or "largest common geographic units." Note that although the number of possible combinations of attribute values is large, most of these combinations will not exist, because of the inherent spatial correlations among attributes. This approach requires significant GIS computation, especially if carried out through true polygon overlay. Raster overlays might be more efficient and adequate, although this would have to be determined on the basis of experience with real data sets.

A database to support land supply monitoring through inventory control must address the problems of homogeneity in regard to attributes and translation across units of accounting. These two problems confound each other. Using only one unit of accounting increases heterogeneity with respect to attributes; sustaining homogeneity with respect to attributes requires using multiple units of accounting. We focus here on describing two possible strategies, which we will call Dynamic Parcels (DP) and Multiple Spatial Units (MSU). We then compare these prototypes to RLIS and suggest that the Multiple Spatial Units approach is likely to be more practical in most situations.

Dynamic Parcels

The Dynamic Parcels approach uses the tax-lot parcel as the basic unit of inventory through all stages of the land development process. Governments managing land can alter stocks of land in various stages of development by making investments or by enacting regulations. The idea of working with parcels as the only spatial unit is initially attractive because of its seeming intertemporal simplicity. The stock of land with any set of attributes is the set of parcels with those attributes. The inventory of parcels in each state is maintained over time by recording transition events. Transition events, such as infrastructure provision, "move" parcels from one inventory to another by changing the values for the attributes that define separate inventories. A "reorder" for a particular inventory consists of moving parcels from the stock in one inventory to the stock in another, by changing the value of an attribute. Note that parcels may have additional attributes, such as size of parcel, that do not distinguish between inventories but are necessary in analyzing supply.

Yet it is a fantasy to think that a strictly parcel-based land monitoring system requires only that we keep track of the status of each parcel as it moves through the sequence of inventories in the development process. Our own attempts to implement such a database highlighted the two problems that plague attempts to implement parcel-based land supply inventories. First, attributes pertinent to determining the inventory status of land are not necessarily homogeneous throughout a parcel. A tax-lot parcel may be partly in a floodplain and, thus, only partly available for development. A large parcel may be zoned partly for commercial use and partly for residential use. Even a relatively small parcel may be under development as a dense town house project, but only partially completed. Plan designations and infrastructure availability may also not be homogeneous across large parcels, particularly in areas that are not yet developed for urban use. Second, parcels are dynamic. That is, parcel geography changes. Changes occur most frequently at the urban fringe following residential subdivision, and in areas of major redevelopment following land assembly and replatting. Subdivision will create street rights-of-way, which must be accounted for as well. Thus, even if we use parcels as the accounting units throughout, they will not be the same parcels.

There are several ways to handle parcels that are heterogeneous with respect to attributes. Where a value of an attribute applies to only part of a parcel, it may be described as a proportion of the parcel, as existing in the parcel, or as the dominant value for that attribute for that parcel. If, for example, most of a parcel is zoned for single-family residential, the zoning attribute for that parcel would be single-family residential. In any case, we would be unable to divide these parcels into subparcels and determine appropriate attribute values for the subparcels, because we would not know the geographic pattern within the parent parcel. This approach would require that combinations, such as "in a floodplain" and "not serviced," be determined externally before coding in order to calculate deductions from available area correctly. These combinations could be determined with the use of GIS operations or by direct coding. The choice would depend on the resolution and timeliness of GIS coverages versus the inherent logic of the combinations. An example of the latter is that all floodplain land could be treated as unserviced, so that the question would be whether any of the rest of the parcel was also unserviced.

Partial development could be estimated by comparing attributes from the building permit to attributes of the parcel and regulations applying to the parcel. It would be difficult, however, to distinguish between "underdevelopment," defined as a parcel built at less than the maximum density, and "incomplete development," defined as a parcel on which some building has occurred with more to occur later. If the density for the entire parcel is less than the minimum required density, it could be flagged as incomplete, but this might apply to only a small portion of relevant cases, because minimum density zoning is so new. Whether this aspect can be handled through a set of computerized rules, or will require some efficient form of human intervention, can be determined only by experiment. Potential for redevelopment might be determined similarly, perhaps using the ratio of improvement value to land value, as is currently done in some applications in Oregon.

The preceding discussion of this approach presumes that the definition of a given parcel remains constant over time so that transition events can be applied to a parcel consistently. To determine what land is "upstream" of building permit approval (in any state of the development process prior to building permit), for example, we must be able to identify all land that is within a UGB, zoned, and serviced and either subdivided or not subdivided. We must therefore know what parcels exist at a particular time. If we rely on transition event coding rather than snapshots of the state of the entire system, transition from not subdivided to subdivided parcels might be encoded based on the subdivision approval process.

Multiple Spatial Units

An alternative approach to organizing data for inventories is to start from the inherent entities of the data at different stages in the development process and consider whether these can be linked through time more easily than parcels can. The initial attraction of this approach is that it follows the standard principles of database design by defining entities for which attributes are inherently homogeneous. The different attributes that are pertinent at various stages of development apply to different spatial units. For example, attributes that describe regulations and infrastructure services apply to zoning boundaries and service-shed boundaries, respectively. Land use types usually apply to dwelling units or building square feet. The Multiple Spatial Units approach uses whatever spatial units are HGAs at any stage in the development process.

This approach avoids the major difficulty that arises in tracking the dynamic transition of parcels through the subdivision process. Prior to subdivision, the inventory is kept in units of land area (e.g., acres) without regard to parcels. Interpretations of available supply are made in gross terms that include areas for streets and auxiliary land uses such as schools and parks. Estimates of the number of dwelling units that can be realized from a given stock of land are based on zoning and infrastructure capacity for each land type and are measured in terms of gross land area. For land accounted for in attributes of parcels, estimates should be based on net acres rather than gross acres. For single-family residential, the number of lots could be used. Once building permits are issued, the estimates can be based on number of dwelling units for residential or occupiable square feet for nonresidential.

Once land is subdivided, especially in the case of single-family residential lots less than a quarter acre in size, these parcels tend to be HGAs with respect to pertinent attributes. Larger parcels are likely to be disaggregated into derived HGAs not corresponding to parcels, because the time of construction is likely to be different among lots. Most quarter-acre or smaller lots will not be heterogeneous with respect to regulations or other pertinent attributes for developed land. Thus, because it uses the geographic entities to which attributes inherently apply, the MSU approach is likely to become equivalent to the parcel approach after subdivision occurs. Even for nonresidential parcels and multiple-family parcels, the site plan approval "event" will yield timing attributes specific to the parcel, and thus will "flag" a predictable point in the development process at which the appropriate spatial unit will likely become a tax lot. At this stage in the development process, the parcel geography is likely to be stable and rights-of-way will have been excluded. After subdivision, therefore, inventories can be tracked in terms of parcels, thereby allowing the planner to take advantage of the significant investment in parcel databases for inventory management of these stages of the development process.

Implementing Dynamic Parcels or Multiple Spatial Units: A Comparison

The Dynamic Parcels (DP) approach tries to simplify the task of translating between spatial units of accounting by always using parcels, but this tends to increase heterogeneity and thus complicate the passing of attributes. The Multiple Spatial Units (MSU) approach simplifies the passing of attributes by using whatever units of accounting are homogeneous, but thereby complicates the translation between accounting units. General experience with such databases suggests that the MSU approach will be more efficient, but conclusive evidence can be gained only through experience in setting up and using operational land supply monitoring systems. These trade-offs depend on the handling of transition data, which is the starting point for comparison.

Either approach relies on the same data sources for the events identifying transitions between states. We can use subdivision approvals to indicate transition from serviced land to lots. Subdivision is associated with initial and resulting parcels, but often in a complicated way. Subdivision approval might be treated as the vesting of development capacity in particular land parcels. An equivalent must be found for industrial and commercial land and multifamily complexes that may not be subdivided. We can use building permits to indicate the transition from land that is subdivided and serviced, to land on which construction will be occurring. Building permits are likely to be associated with a parcel, at least for new construction, but the permitted construction may use only part of a parcel. We can record occupancy permits as indicators of building completion and thus mark the transition from projects under construction to those available for use. Occupancy permits generally apply to dwelling units or portions of commercial or industrial buildings that can be measured in available square feet. They may also be associated with street addresses rather than land parcels. Thus, these data about the succession of events that transform land from one state to another in the sequence of inventories apply to land parcels, construction projects, and occupancy units (dwelling units or square feet), respectively. These data are not all inherently attributes of parcels. Thus, the transition events are not all transitions of parcels, and they do not necessarily apply to parcels homogeneously.

Inventory control models can be computed from either database approach with differing degrees of difficulty and different expectations about data validity and reliability. It is possible to compute the available supply upstream of any stage in the development process by estimating in dwelling units, for example, the stock in each preceding stage and summing dwelling units for these stages. It is thus possible to structure an inventory model relative to the stock in any stage. Given explicit and fixed formulas for each of the conversions to dwelling units for given land types, it is always possible to identify the number of dwelling units in a stage and to express events that supply or delete units from that stage in terms of dwelling units. A conversion from a regulated and serviced land type to subdivided parcels would remove that number of acres from the HGA of that land type and, thus, that number of dwelling units associated with those acres. It would add a number of dwelling units to the subdivided land equal to the number of lots of subdivided land. Similar conversions would apply for the DP approach. Thus, the two approaches are approximately equivalent in this regard.[5]

The DP approach appears to provide for simpler transitions between inventory states, especially at the urban fringe, because the spatial unit remains constant, but tax-lot parcels change with urbanization. The MSU approach should simplify changes over time in infrastructure service areas and regulations. It should be easier to update the set of HGAs than it would be to associate these changes with a set of changing parcels. MSUs avoid most of the problems of keeping track of changing parcels and decreases the number of geographic units that must be updated when such changes occur. These updates would, however, still involve true polygon overlays, with the various error and computational burdens that imposes. Yet we would expect this to be easier than keeping track of parcel changes, because the amount of line work necessary to encode service areas or regulatory areas should be much less than the amount of line work needed to record changes in individual parcels. In many cases, updating will require no new geography, for example, changing "infrastructure planned," to "infrastructure programmed," to "infrastructure available." We cannot be certain which approach will be simpler to implement in regard to transitions of spatial units without experimenting with actual implementations.

The MSU approach avoids the problem of attribute heterogeneity because attributes of HGAs can be attributed uniquely to subdivided lots. HGAs are by definition homogeneous with respect to all attributes. Even after subdivision and urbanization some large parcels may not be homogeneous with respect to all attributes of interest. These parcels could be maintained as derived HGAs. Single-family parcels, for which this should not be necessary, will constitute the largest portion of urban built-up area as well as the largest portion of line work because each individual parcel is so small. Thus, even if the savings in accounting for transitions of spatial units occur only for single-family parcels, they should be significant.

Each approach raises particular questions of implementation that should also be considered. At any one time some land may be at each stage of the development process. We must, therefore, consider spatial units that are at different stages in the development process and, thus, for MSU, different types of spatial units. It should be possible to maintain reliable distinctions between land accounted for in derived HGAs and land accounted for explicitly in parcels. For any point in time, the derived HGA geography and the parcel geography would be maintained as part of the same coverage, thus ensuring

---

[5] The approaches to data structures for Dynamic Parcels (DP) and Multiple Spatial Units (MSU) are compared in Table 3.3. The crucial difference is that DP uses tax-lot parcels as the spatial unit through all stages of development, whereas MSU uses inherent or derived HGAs as its spatial units for accounting. In general, MSU uses derived HGAs from intersections of the spatial geography of combinations of attributes for stages of development prior to subdivision, tax-lot parcels after subdivision, and dwelling units or square feet after occupancy.

TABLE 3.3 Comparison of Data Structures

	Multiple Spatial Units		Dynamic Parcels	
Development Stage	Spatial Unit	Unit of Measure	Spatial Unit	Unit of Measure
Potential Urban Land	Derived HGA	Gross acres	Tax lot	Gross acres
Urban Buildable	Derived HGA	Gross acres	Tax lot	Gross acres
Serviced	Derived HGA	Gross acres	Tax lot	Gross acres
Subdivided	Tax lot	Net acres	Tax lot	Net acres
Building Permit	Tax lot	DU or sq ft	Tax lot	DU or sq ft
Occupancy Permit	DU or sq ft	DU or sq ft	Tax lot	DU or sq ft

Notes:

HGA = Homogeneous Geographic Area; DU = dwelling unit; sq ft = square feet of available building space.

Conversions of units from gross acres to net acres, from gross acres to dwelling units or square feet, and from net acres to dwelling units or square feet are discussed in the text. Dwelling units are used for residential, and square feet for all other land uses.

consistency of topology and boundaries. Using such a coverage confirms the logic that the parcels are used if and when they are HGAs, which is generally the case after the land is subdivided. Working from this coverage as it exists at any point in database time, we would retrieve data for HGAs, some of which would be parcels.

There is no possibility of confusion as to whether parcel data are being retrieved for both a parent and a descendant parcel, because there will be no parents in the coverage if their descendants are in the coverage. Thus, for example, even if there are attribute data already in the tabular database for new parcels that have not yet been added to the coverage, the attributes of these parcels will not be retrieved, because the identifiers of these parcels will not be in the coverage. The attributes of the HGAs in that area will be retrieved instead, which may be less up-to-date in world time, but will not confound operations in database time. In contrast, a DP database structure must keep the parent parcels in order to track change over time and, thus, might retrieve a parent parcel that happened to have the same parcel identifier as a descendant parcel. Or a parent parcel might be treated as subdivided and thus reported using measures net of streets rather than gross area. In the DP approach, greater care would be required in keeping attribute data and spatial coverages synchronized and specific flags would be required to identify whether a parcel should be treated as gross or net area.

Working with the inherent entities of transition events and derived HGAs, as in the Multiple Spatial Units approach, appears to have significant advantages over the Dynamic Parcel approach. This expectation is based on the presumption that once parcels are subdivided, there will be very little further subdivision activity and that parcels will be stable from that point. This should be true if we are able to keep land that is not fully subdivided in the derived HGA domain. For example, in a phased subdivision, a portion of the parent parcels may be subdivided into lots and the rest held as one or a few large parcels. These remaining parcels should still be treated as part of the derived HGA domain, not the subdivided parcel domain. It should also be relatively easy to record parcel changes and maintain pointers in redevelopment areas and other instances of parcel changes after subdivision simply because the number of cases should be small.

RLIS AND METRO'S CAPACITY ANALYSIS REEXAMINED

With the aforementioned frameworks as points of reference, we can now evaluate the data structure of RLIS and its application for land supply monitoring. We begin with an analysis of the database structure, followed by an analysis of Metro's recent efforts at land supply and capacity monitoring.

Database Structure

A quick review of Table 3.1 reveals that Metro currently collects almost all of the data attributes needed for comprehensive land inventory management: growth boundary status, building permits (yielding existing use, density, estimate of infrastructure demand), and zoning. Only in rare instances, however, are historical data kept when parcels are updated. Even these historical data are extremely difficult to use, because the intertemporal relationships are not maintained in a practically useful manner. Another issue, which we will not address here, is data consistency across data collected by different local jurisdictions and agencies.

RLIS includes a complete parcel data layer, but it is not structured as a purely Dynamic Parcels approach. With the exception of those attributes maintained by the tax assessor's office, the attribute data are not kept as parcel attributes and some of the layers are not fully consistent with the parcel base map. The polygons in the floodplain coverage, for example, reflect estimates derived from contour lines rather than parcel boundaries. The zoning and plan designation coverages are also based on polygons that may intersect parcel boundaries.

Although the structure of RLIS is not fully developed either as a DP or as an MSU database, it does contain data that could be used to create either. Our experience with using RLIS as the basis for implementing a DP approach (Knaap et al. 1996) illustrates several major problems related to parcel heterogeneity and dynamics.

Creating parcel attributes by overlaying various coverages on the parcel coverage is a feasible but messy process, as described earlier. For this reason, Metro assigns plan designations to individual parcels, using the plan designation of the polygon that contains the parcel centroid.

The problem of changing parcel boundaries is also difficult to overcome in RLIS. RLIS is modified to delete parcels that no longer exist and add parcels and their attributes as they are created. The new parcels are not, however, numbered or otherwise identified in a way that is sufficient for reliable identification of parent and descendant parcels. The parent parcel identifier number may be reused for one of the descendant parcels. The parent parcel boundaries may not be congruent with the aggregate of the descendant parcel boundaries. Sometimes two or more parent parcels are combined, then subdivided so that a descendant parcel may or may not have been derived from one or both parent parcels. Conceptually, a database structure can be designed that combines all the parents and then, with no elapsed time, subdivides them with pointers to each descendant; practically, however, the task is extremely difficult and imprecise.

The persistence of the changing parcel boundary problem in RLIS imposes limitations on the use of parcel data for the temporal tracking and analysis of land supply necessary to manage a continuous inventory. These data cannot be used (1) to determine an inventory status at any arbitrary time, (2) to determine rates of change over time in parcel status, or (3) to pass attributes from parent parcels to descendant parcels. Up-to-date and time-consistent parcel data may not be available in a timely fashion, because the RLIS update process is manual. Problems conflating real- or world-event time with database-event time are thus likely to exacerbate temporal data management.

Land Supply Monitoring

If the primary purpose of developing a land information system is to monitor land supply and facilitate more effective growth management, then it is not clear that a Dynamic Parcel approach will be cost-effective. Metro's procedure for estimating a snapshot of developable lands and to analyze development capacity suggests that the MSU approach is likely to be feasible for implementing an inventory control model (Portland Metro 1997). Metro's method begins with the vacant land coverage, which is based on HGAs. These HGAs may be coded directly from aerial photographs as areas of undeveloped land or derived from combinations of other GIS coverages and aerial photo interpretation. From the vacant land coverage, Metro subtracts vacant but platted single-family land (based on parcels), environmentally constrained land (based on appropriate HGAs), and land needed for parks and infrastructure (based on population ratios). From the resulting HGAs, the remaining vacant and developable land is further divided by comprehensive plan designations, which are the basis for calculating capacity. The capacity of single-family land that was already platted is added to the calculated capacity. A parcel coverage is used to identify the development capacity of platted (subdivided) vacant lots. This process uses HGAs, including parcels in the few cases in which they are the appropriate HGA for a particular set of attributes.

The primary limitation of Metro's land supply and capacity analysis is that it is conducted as a static exercise. That is, instead of analyzing how much land is currently available, estimating how long this stock of land will last, and determining when the inventory should be expanded, Metro's analyses focus on whether there is sufficient inventory to meet projected growth for an arbitrarily chosen 20-year period. Hence, procedures prescribed by state law require Metro to take an approach different from the well-established principles of inventory management (Portland Metro 1997).

IMPLICATIONS FOR FUTURE WORK

To test our arguments, the perpetual inventory method should be implemented, using the Multiple Spatial Units approach. This trial must be of sufficient size, in both time and space, to discover whether the presumptions about the character of HGAs and parcels and the frequencies of parcel change are correct. Expectations about the relative efficiency and validity, in practice, of the data collection and database operations will require further analysis based on empirical data.

Our experience with RLIS suggests that MSUs are better suited as units for structuring a database to support an inventory approach to growth management. More important, the inventory control approach, in contrast to the snapshot land supply approach, has significant potential to improve the ability of current growth management tools to achieve their objectives. In particular, experience with UGBs suggests that they are blunt instruments for both infrastructure timing and land use control. Current practice in assessing available land focuses on snapshots, which are often two years out of date and poorly suited to compute trends, rates of change, or specific inventory control strategies. Data structured to support an inventory management approach should allow planners to increase the precision of UGBs with respect to infrastructure timing, without unintended consequences for housing prices or economic development.

REFERENCES

Ding, Chengri, Lewis Hopkins, and Gerrit Knaap. 1997. Does planning matter? Visual examination of urban development events. *Landlines* 1: 4–5.

Godschalk, David R., Scott A. Bollens, John S. Hekman, and Mike E. Miles. 1986. *Land supply monitoring: A guide for improving public and private urban development decisions*. Boston: Oelgeschlager, Gunn & Hain, in association with the Lincoln Institute of Land Policy.

Greene, J. H. 1969. *Production and inventory control*. New York: McGraw Hill.

Haynes, K. E., A. Krmenec, D. Whittington, W. F. J. Eichelberger, and T. D. Georgianna. 1984. Planning for capacity expansion: Stochastic process and game theoretic approaches. *Socioeconomic Planning Science* 3: 195–205.

Hopkins, Lewis D. 1977. Methods for generating land suitability maps: A comparative evaluation. *Journal of the American Institute of Planners* 4: 386–400.

Intriligator, M. D., and E. Sheshinski. 1986. Toward a theory of planning. In *Social choice and public decision making,* W. Heller, R. Starr, and D. Starrett, eds. Cambridge, U.K.: Cambridge University Press.

Knaap, Gerrit J., Lewis D. Hopkins, and Kieran P. Donaghy. 1998. Do plans matter? A framework for examining the logic and effects of land use planning. *Journal of Planning Education and Research* 1: 25–34.

Knaap, Gerrit J., Lewis D. Hopkins, and Arun Pant. 1996. Does transportation planning matter? Explorations into the effects of planned transportation infrastructure on real estate sales, land values, building permits, and development sequence. Working paper. Cambridge, Mass.: Lincoln Institute of Land Policy.

Portland Metro. 1997. *Urban growth report: Final draft.* Portland, Oreg.: Metro.

Segoe, L. 1941. *Local planning administration.* Chicago: International City Management Association.

Sippen, Daniel, and Robert L. Bulfin Jr. 1997. *Production: Planning, control and integration.* New York: McGraw-Hill.

COMMENTARY: HOMOGENEOUS GEOGRAPHIC AREAS AND PARCEL-BASED GIS

Scott A. Bollens

This chapter challenges the common acceptance of parcel-based databases as the preferred foundation for land monitoring systems, stressing that data collection and analysis should be done in ways that support planning and growth management policy-making. In addition, it presents an inventory approach to tracking land supply that contrasts with the commonly used "snapshot" approach.

Attendant problems in establishing parcel-based data for land monitoring include (1) changing parcel geography and difficulties in intertemporal monitoring when parcel boundaries change as a result of subdivision and (2) the heterogeneity of subparcel activity (partial use of a parcel, multiple uses). If longitudinal tracking, which is inventory-based monitoring and not snapshot monitoring, is important to planners, then these problems (especially the first) seem to argue for the use of different units of analysis. The difficulties related to changing parcel geography are analogous to the difficulties associated with tracking census data over time and changing census geography (such as tract boundaries).

Hopkins and Knaap propose the use of Homogeneous Geographic Areas (HGAs) to analyze the characteristics of the land supply. HGAs are generated through an overlay of nonparcel coverages representing land classified according to such factors as regulatory status, environmental factors, and infrastructure availability. Each HGA supply category that would result from this overlaying has similar values across land characteristics (for example, all land in one category has the same zoning, environmental quality, and degree of existing or planned infrastructure). HGAs provide greater stability of spatial unit boundaries than tax parcels and minimize the heterogeneity problem because HGAs, by definition, are products of homogeneous characteristics. This presents an alternative spatial accounting approach for urban fringe land that would enhance temporal monitoring by tracking land in different supply categories by gross acreage, and not artificially forcing this inventory into a parcel-based geography. The proposed system combines the use of HGAs as the units of analysis before fringe-area land is subdivided, and parcels as units of land tracking after subdivision. This is so because, after subdivision, parcel boundaries become more stable and reliable for temporal monitoring. The Portland RLIS case, as described, appears to be a limited HGA approach, incorporating just one or two nonparcel layers to define the homogeneous polygons.

The HGA approach outlined by the authors helps us contemplate the strengths and weaknesses of the tax lot or parcel as a spatial unit for data collection and analysis. In systems that include among their multiple coverages a parcel base map, planners are able to identify finer patterns of land supply and development activity. There is also flexibility of aggregation, as planners collect data at the finest level of detail initially and thus are not forced to a coarser level of resolution at the start. This is consistent with a rule in social science: Always keep data at its most specific level at input stage. However, in parcel-based systems there are problems with longitudinal tracking (as noted), as well as high costs and potentially burdensome updating requirements. Thus, even if the unit of collection for some data is the parcel, land supply monitoring needed to ensure effective growth management may require that planners use other units of analysis or evaluation that are more appropriate for temporal tracking. If HGAs are more fruitful for monitoring and analysis for urban and regional planning, then there would appear to be distinct advantages to moving in that direction and detaching from the land management and accounting biases often inherent in parcel-based and assessor's data.

On the other hand, the use of HGAs to track dynamic aspects of land supply may cause unsuspected difficulty and will require careful attention. First, two of the three characteristics the authors suggest that would demarcate HGAs (namely, regulatory status and infrastructure availability) are themselves susceptible to change. Thus, the boundaries of HGAs may not be stable, moderating somewhat their capacity to help planners track land supply over time. Second, because many environmental features (such as floodplains) do not match local municipal boundaries, HGAs may, in particular parts of the urban region, be multijurisdictional. If a growth management system does not have a strong regional component, then data from cross-jurisdictional HGAs will be inappropriate for municipal decision makers and planners. Third, more clarification is needed regarding how presubdivision HGA databases and postsubdivision parcel databases can be represented to paint a composite picture of land, development, and the dynamics of both, for any particular part of the urban region.

The chapter also proposes an alternative method for monitoring land and rightfully stresses the importance of utilizing an ongoing inventory approach to land supply management. It highlights the collection and analysis of data that respond to planners' needs to effectively and accurately track changing land and development conditions. In its proposed spatial approach utilizing both HGAs and parcels, and through its case study of the Portland RLIS, the chapter also shows that the future of land monitoring systems lies not in an "either-or" choice. Rather, planners and data managers should seek effective and efficient approaches that combine the temporal monitoring benefits of HGA tracking with the detailed richness of parcel-specific accounting.

COMMENTARY: PUBLIC VERSUS PRIVATE INTERESTS IN LAND MONITORING

George R. Rolfe

One of the most challenging issues facing land monitoring is the difference in the objectives for and interpretations of land information that exists between public and private interests. Public-sector users of land use data can generally be grouped in two broad categories: those charged with long-term, comprehensive planning and those addressing short-term project approval and permitting. Private-sector users of land use data are almost exclusively focused on short-time-frame project development and marketing activities. In addressing these differing perspectives, I draw on my experience as a former developer of both public and private real estate, and 12 years of teaching real estate development. Having also served on the King County Growth Management Land Capacity Task Force, I have seen, firsthand, some of these issues being worked out.

Hopkins and Knaap identify at least two critical and frequently contentious issues related to the use of GIS by local governments in their efforts to plan and manage land use and growth. The first is how to determine the appropriate vocabulary necessary to convey useful information between governmental employees and elected officials, on one hand, and the primarily private development community, on the other. The second issue concerns the differences between the time frames typical of public policy planning and private project decision making. Both issues have technical implications for the collection and management of land use data.

An Appropriate Land Use Inventory Vocabulary

To address the issue of vocabulary, the authors propose a two-tiered system of parcels and Homogeneous Geographic Areas (HGAs). From a government planning perspective, HGAs make sense because they allow for the relatively easy conversion of raw land amounts into units of development. This is useful for making long-range projections of development potential, particularly at the fringe of urban areas, and for assessing the appropriateness of Urban Growth Boundaries (UGBs). Further, HGAs avoid the problem of distinguishing between the development potential of unsubdivided parcels that contain land *currently* suitable for development, land *ultimately* suitable for development but not yet serviced, or land that is *unlikely ever to be developed*. Thus, inventories of land based on common development characteristics rather than historical accident of ownership pattern yield more accurate estimates of gross holding capacity (e.g., dwelling units, square feet of built commercial space) than those yielded by most parcel-based inventories.

A current private-sector criticism of parcel-based land supply monitoring systems concerns whether owners of individual parcels will make land available at current market prices and, if not, whether those parcels should be included in the inventory. Although not intended for this purpose, HGAs are an interesting way of avoiding this issue early in the public process, inasmuch as the connection between HGAs and individual parcels need not be made before subdivision and platting occur. When parcels are platted and removed from the HGA inventory, however, "market factors" would have to be applied. These could be based on observed market behavior rather than on intuitive judgment.

Differentiating between land inventoried in HGAs and land in parcels may also serve to help the private development community focus on the short-time-frame parcel-based inventory. This should have the practical effect of separating policy discussions about long-range planning based on projected holding capacity in HGAs, from the more contentious issues associated with the permitting of specific parcels. Current debates about the long-term supply of land, in fact, act as a surrogate for concerns about the immediate availability and price of parcels of developable land.

Private developers customarily think in terms of parcels first and dwelling units second. They look for available land suitable for development and then calculate "yield" in terms of development potential to validate the price of the land. In fact, most developable land on the market today is ultimately priced by its permitted development potential—mitigated by issues of location, physical characteristics, access, and surroundings—and not by its size. Thus, accurate estimates of capacity are more important to the private sector than simple inventories of "developable" lands.

Land Price Versus Supply   Should land inventory systems be based primarily on price instead of supply? Current monetary policy at the federal level suggests this to be the case. However, land market efficiency must be understood with respect to conditions that differ from those of stock markets. Information on price and availability is not uniformly available to all buyers and sellers on a real-time basis. Further, information about land price is not easily verified as to its accuracy, nor are there accurate data to support projections of the land's economic use potential, which would be required to make efficient adjustments in price. Finally, there is no common framework within which all interested buyers and sellers can exchange the information that does exist. Nevertheless, although the current land market is not efficient in terms of classical market theory (North 1990), there is an "installed base" of buyers, sellers, brokers, and other service providers familiar with current market practices. This gives existing participants a clear advantage when dealing with either government policymakers or new entrants to the land market. Government is not currently seen as a major participant in the land market, and therefore its entry into the collection and management of data is unlikely to be seen by the private sector as having a significant impact on land markets. If, as Hopkins and Knaap suggest, local government parcel databases created for policy purposes could also become effective sources of market information, then these databases could compete with existing private data sources to inform market behavior.

Planning Data Versus Market Data   The proposed Multiple Spatial Unit approach integrates HGA-based data with existing public and private parcel databases. Many data relating to government permits are tracked on a parcel basis, and often incorporated in a GIS. Data transfers to the land monitoring system could be automatic once entered in the issuing agency's database. As newly platted land is added to the parcel-based inventory, and as new building or occupancy permits are issued, appropriate adjustments to capacity could be made in the parcel inventory. Data on private sales activity, including price, are also tracked on a parcel basis. Government permitting and private sales data could therefore be linked. Thus, a two-tiered system, clearly separating planning data from market data, is likely to be seen by private development interests as an improvement to current land inventory processes.

Potential Impediments to Implementation   Several factors present potential difficulties to the implementation of the two-tiered system proposed by the authors. First, current land monitoring systems rely heavily on parcels as the basic unit of inventory. Because of an "installed base" of hardware, software, and personnel to support parcel-based land monitoring, organizational opposition to "retooling" to accommodate HGAs is likely. However, GIS used for land monitoring is, relatively, in its infancy, and may be influenced by the successful implementation of a prototype system combining HGAs and parcels. Second, there are significant legal and institutional barriers to system integration. Confidentiality impedes data sharing, both between public and private sectors as well as among public agencies. Skeptical private-sector attitudes toward public-sector data (particularly assessors' data) are likely to create further obstacles to integration. Third, the HGA approach is most suited to monitoring land supply for residential uses, particularly single-family development. Application of HGAs to commercial lands must resolve the technical problems of relating land area consumed to floor area of new commercial development.

Disparate Decision-Making Processes and Time Frames

Hopkins and Knaap do not explicitly address the issues arising as a result of the discrepancy between the long time frame used in planning and the short time frame used for permitting decisions. However, their perpetual, multiple-spatial-unit-based inventory control approach could very well address the needs of both. Local governments typically plan and set policies for time periods ranging between 5 and 20 years. This long time frame of reference is also a fundamental tenet of traditional planning theory, as evidenced by the use of terms such as "long-range," "comprehensive," and "master" plans. These are terms that are seldom used by private decision makers.

Set against this long time frame for planning and public policy, most local governmental decisions regarding individual permits for development projects occur within a time frame ranging from three months to several years. Often the two functions—long-range planning/policy determination and development permitting—are performed by two separate entities within local government. The permitting process is of greater interest to private development interests.

Beyond operating within different time frames, the public and private sectors also differ in the ways in which they make decisions. Over the past 20 to 30 years, development permitting has become largely driven by "process" rather than substantive "content." Although planning has traditionally been concerned with substance, recent evidence points to "long-range comprehensive" planning efforts becoming driven more by process than by content (Helling 1998). A possible explanation for this growing focus on process can be found in the way public decisions are made. Public decisions are linear processes, with required steps often spelled out in detail before the process begins. These steps cannot be easily modified according to specific outcomes. The eventual decision is, most often, the last step in the process and is not known (or, at least, not publicly acknowledged) prior to completion of the process. This leads to a long, labor-intensive, and outcome-uncertain process, one that is often politically contentious. Consequently, once public decisions are made and policy is adopted, there is a reluctance to revisit old issues even when new information indicates the wisdom of doing so.

In contrast, the private decision-making process has many opposite characteristics. It is primarily outcome oriented. Seldom is a question addressed that does not have some level of private commitment to a specific answer or decision before the process is undertaken. Although private decision makers do modify their positions based on new information and conditions, their decision-making process is driven by a fundamental intention to arrive at a predetermined conclusion. Procedures are mostly informal, seldom written down, and easily modified if outcomes are not satisfactory. Content is what matters to the private decision maker. Typically, a developer wants to know whether a permit will ultimately be issued at the end of the process. On the other hand, the permit issuer's role is to ensure correct "protocol" (in this case, to ensure that the correct process is followed) rather than to commit to a specific outcome. These conflicting purposes underlie the mistrust that exists between government and private interests.

Recent Changes in Time Frames for Private Market-Related Decision Making   There is a need to reconcile differences in time frames and decision-making processes within which public and private decisions are made. The length of time needed to make public decisions has tended to increase over the past 30 years, whereas market cycles for development have tended to become both shorter in duration and more fragmented. Throughout the period from 1945 through much of the 1960s, real estate markets were need-driven, resulting in relatively stable cycles of demand based on macroeconomic issues tied to population growth. In contrast, current markets are driven by consumer wants. Most people who *need* housing or other real estate space are often unable to compete economically for available new products, leaving the new product real estate market to address the *wants* of those individuals and firms with the economic capital to satisfy those wants. This has led to a much less predictable market and to a dramatic shortening of the time horizon within which development decisions must be made. This shortening of the time horizon is typical of all private aspects of the real estate development process, including finance markets, rental and sale markets, and the markets for materials and services necessary to develop and build real estate.

As public and private time frames have diverged, data useful to local governments in making land use plans and policy have become increasingly irrelevant to private developers. As public decision making has become more process-driven and lengthy, private developers have become frustrated with government's inability (and perceived disinterest) to deal with the "market realities" experienced by private decision makers.

Implications of the Proposed Land Inventory Process   Differences between public and private time frames have clear implications for current public debates over UGBs. Basically, government is concerned with land available to accommodate long-term growth within a defined UGB, and private development interests are focused on the supply of buildable lots available within short time-frame market opportunities. This dichotomy may be effectively dealt with in the proposed perpetual inventory control process. This inventory process makes it theoretically possible to adjust Urban Growth Boundaries based on real-time data and, hence, on a time frame more consistent with market forces. The question is whether local governments have the resources and political will to tackle such an adjustment on the three- to five-year basis implied by this chapter.

The perpetual inventory could benefit private developers by focusing on how public policy should compensate for current market inefficiencies—one of the most contentious aspects of current policy debates. A revolving, real-time market inventory model based on price indicators can determine when a land inventory is insufficient, and help trigger appropriate public decisions affecting zoning and provision of public infrastructure to increase land capacity. Such adjustments should, theoretically, make the land market more efficient and, at least indirectly, affect the price of available land for development. A second benefit, which the private development community should welcome, is the availability of verifiable "market" data—both making public policy more responsive to market forces and providing an "impartial" database of parcels as they move through the public permitting process.

However, given the present state of mistrust between government and the private sector, it is possible that such a publicly managed database, even though accurate, current, and widely available, may not be credible. An intermediate step may be to have such a parcel database managed by an independent third party with some level of recognized impartiality. A good case might be made for universities as the managers of land supply data; for example, the University of Texas has successfully served in this capacity for the Dallas-Fort Worth metropolitan area.* Local governments would presumably be willing to share their existing extensive data on permitting with university personnel. In addition, it is conceivable that existing private data providers would be willing to share their market-oriented data, making the parcel database that much more useful to both public and private decision makers. The issue of credible management would likely become critical, should a given local government choose to implement Hopkins and Knaap's land inventory proposal.

References

Helling, Amy. 1998. Collaborative visioning: Proceed with caution! Results from evaluating Atlanta's Vision 2020 Project. *Journal of the American Planning Association* 3: 335–349.

North, Douglas. 1990. *Institutions, institutional change, and economic performance.* Cambridge, U.K.: Cambridge University Press.

---

* Similarly, the University of South Carolina Humanities and Social Science Computing Laboratory maintains files for all localities in the state.

4

Montgomery County, Maryland: A Pioneer in Land Supply Monitoring

David R. Godschalk

Montgomery County, Maryland, has pioneered the application of land supply monitoring, using it as a major tool to implement the nationally recognized county growth management program. This chapter presents an update of a previous 1985 study of land supply monitoring in Montgomery County, published along with several other case studies in *Land Supply Monitoring: A Guide for Improving Public and Private Urban Development Decisions* (Godschalk et al. 1986). The 1998 status of the Montgomery County approach to land supply monitoring is reviewed in terms of its relationship to a number of growth management tools, including (1) the General Plan, (2) the Adequate Public Facilities Ordinance (APFO) and its implementation through the setting of Annual Growth Policy Ceilings, and (3) the Master Plan for the Preservation of Agriculture and Rural Open Space and its implementation through the use of Transfer of Development Rights (TDRs).[1] Of interest as well is

---

[1] In practice, the Montgomery County approach to planning, growth management, and land supply monitoring is more complex than described in this chapter. The description here is limited to a general overview of selected elements that are related to land supply monitoring. Those interested in complete details should consult the Montgomery County publications, some of which are cited in the references.

Monitoring Land Supply with Geographic Information Systems, edited by Anne Vernez Moudon and Michael Hubner ISBN 0 471371673 © 2000 John Wiley & Sons, Inc.

the potential transferability of the Montgomery County approach to other jurisdictions.

Also considered is the new Maryland statewide growth management program, the Smart Growth Act of 1997, and its implementation through the use of Priority Funding Areas and a statewide parcel-point geographic information system (GIS) called MdProperty View.[2] The chapter concludes by discussing the implications of the Montgomery County approach and the Maryland Smart Growth Act for other jurisdictions interested in using land supply monitoring for growth management.

LAND SUPPLY AND GROWTH MANAGEMENT IN MONTGOMERY COUNTY

Land supply monitoring in Montgomery County, Maryland,[3] is a major tool for implementing county planning and growth management. Unlike many jurisdictions that use their parcel-based GIS primarily for tax assessment and general property information, Montgomery County actively uses its parcel-based system to guide development in accordance with public policy and adopted plans. It has been a pioneer in the development and application of both growth guidance and land supply monitoring in the United States. Montgomery County is an important source of information on "best practices" in land supply and capacity monitoring.

As they have matured, Montgomery County's growth guidance and land supply monitoring approaches have gained in both policy and technical sophistication and power. They have expanded to apply to a broad array of policy issues and have developed advanced technical capabilities. Montgomery County's approach will be difficult for most other local governments to emulate, however, both because of the sophistication of its local planning and growth management and because of its unique setting in a progressive growth-managing state. Nevertheless, it presents an advanced model of the possibilities for using computerized land supply information systems to carry out public

[2] The Maryland approach to growth management contains a number of related elements not described in this chapter, including a Rural Legacy Program of land and easement purchases to preserve rural areas, Brownfields Cleanup and Revitalization Incentive programs, a Job Creation Tax Credit program, and a Live Near Your Work Demonstration Program. The state has also published Smart Growth options for Maryland's Tributary Strategies (Maryland Office of Planning, 1998). As is the case with Montgomery County, the description of the Maryland Smart Growth program here is limited to a general overview of selected elements related to land supply monitoring.

[3] As one of the counties adjacent to Washington, D.C., Montgomery County has experienced high growth pressures for a number of years. From a population of 805,930 in 1995, the county is expected to reach a total of more than 1,000,000 residents by the year 2020 (Maryland State Office of Planning 1997a). The county's 492 square miles encompass both inner- and outer-ring suburban communities as well as rural areas.

policy; and some elements of its approach, as well as the underlying principles, can be transferred to other localities.

Planning Setting

Montgomery County is one of the urbanized "collar" counties surrounding Washington, D.C. It is split by Interstate 270 running to the northwest (Figure 4.1). Its General Plan seeks a harmonious balance of land uses, including a balance between housing and jobs, with the objective that all employees in the county should have the opportunity to live there. In 1990 its jobs and housing were in almost perfect balance, with a ratio of jobs to housing of 1.54, and a ratio of workers per household of 1.55 (MNCPPC 1993, 5).

Montgomery County has a well-deserved reputation as one of the leading progressive planning and growth managing jurisdictions in the United States. Along with adjacent Prince George's County, it is part of the Maryland National Capital Park & Planning Commission (MNCPPC), a bicounty agency created by the General Assembly of Maryland in 1927. The Commission

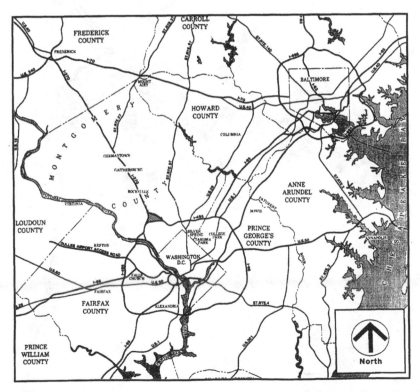

FIGURE 4.1 *Montgomery County and surrounding areas* (*Source:* MNCPPC. 1993. Figure 16, p. 79)

operates in each county through a planning board appointed by the county government.

The 1993 *General Plan Refinement of the Goals and Objectives for Montgomery County* (MNCPPC 1993) amends the adopted MNCPPC 1964 *General Plan,* which introduced the concept of "wedges and corridors" and carried out the regional form vision of the *Year 2000 Plan* for the Washington, D.C., region. The urban corridors follow major transportation routes. The undeveloped green wedges conserve the major natural and agricultural areas (Figure 4.2). The plan is based on a careful and comprehensive land suitability analysis.

The county has actively and effectively implemented its plan through an integrated set of growth management tools:

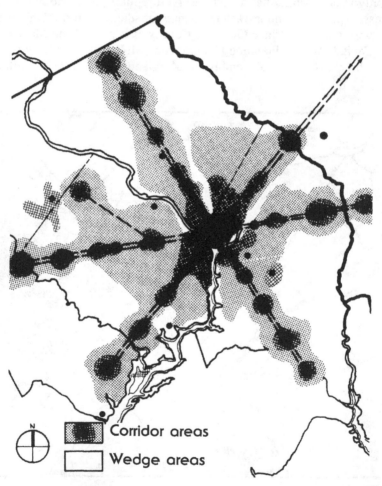

FIGURE 4.2 *Year 2000 wedges and corridors regional concept (Source:* MNCPPC. 1993. Figure 4, p. 8)

- Zoning, used to regulate the location, type, and amount of development. Higher densities are allowed in the corridors, and lower densities are maintained in the wedges, where a functional master plan aims to preserve not only farmland but also the continuation of farming (Figure 4.3) (MNCPPC 1980).
- An Adequate Public Facilities Ordinance, adopted in 1973 as part of the county subdivision ordinance. It is used to coordinate the timing of new development to the capacity of transportation, schools, water and sewer-

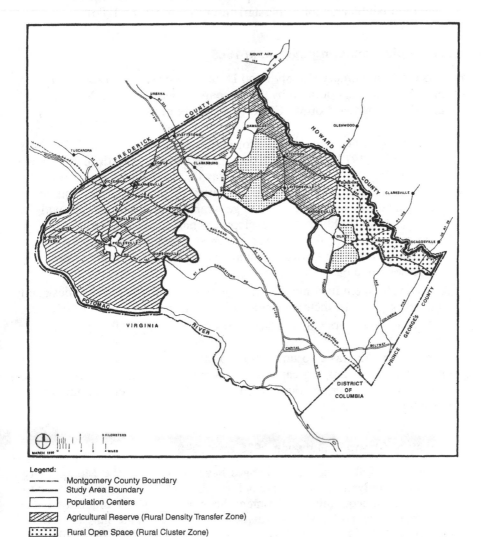

Legend:

————— Montgomery County Boundary
————— Study Area Boundary
☐ Population Centers
▨ Agricultural Reserve (Rural Density Transfer Zone)
⬚ Rural Open Space (Rural Cluster Zone)
▦ Residential (RE-2)

FIGURE 4.3   *Proposed rural area land use and zoning (Source:* MNCPPC. 1980, p. 42)

age, and police, fire, and health service facilities (Chapter 50-35(k), Montgomery County Code 1986).

- An Annual Growth Policy established in 1986. It constrains the amount of development to that which can be accommodated by existing and programmed public facilities within the county's capital improvements program and the state's consolidated transportation program (MNCPPC 1995, 1997).
- A Transfer of Development Rights (TDR) program. It is used to provide an equitable return to owners of property in the wedges, while encouraging the maintenance of viable agriculture there (Pizor 1986).

Land Supply Monitoring System in 1985

In 1985 the Montgomery County Land Data Bank System (LDBS) consisted of three components operated by the Research Division of the Montgomery County Planning Board of MNCPPC (Godschalk et al. 1986):

1. An inventory of land parcel information maintained in the LDBS was initiated in 1977 as part of the Planning Board's growth management accounting system. This inventory was linked to the Adequate Public Facilities Ordinance, in which capacities of roads, sewers, and other critical facilities were allocated to geographic Policy Areas, each with an assigned threshold level of development balanced with facility capacity (Figure 4.4). The inventory was designed to monitor the supply of residentially zoned land, with parcel information coming from the property assessor's database.
2. A Development Information Management System (DIMS) was designed to track projects, including subdivisions, in the development pipeline.
3. A separate system was in place to track building permits.

At that time, work was also under way to integrate the systems and to add parcel and project mapping capabilities, as well as to improve user access. The systems used a combination of mini- and microcomputers running on in-house software.

Overview of Changes Between 1985 and 1998

Since 1985, the following major changes have occurred in the Montgomery County land supply monitoring system. First, the system has been expanded to address issues of economic development. Specifically, an assessment of available commercial and office sites was made and presented in the *Site Characteristics Inventory* (MNCPPC 1998a). A response to concerns about the availability of sites for new employment facilities, this work provides an inventory of vacant and redevelopable commercial and office land, within the context of environmental constraints, growth policy and adequate public facilities limitations, im-

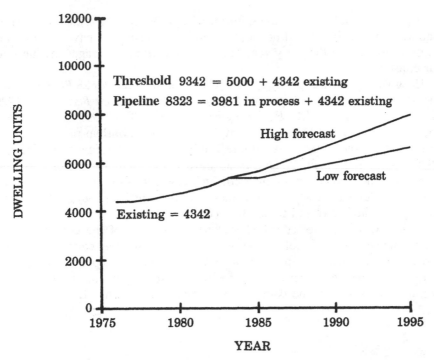

FIGURE 4.4 Example Policy Area development threshold (*Source:* Godschalk et al. 1986, p. 147)

pact taxes, and other factors affecting the ability to develop. The inventory expands knowledge about the county's nonresidential development "far beyond information previously available, due in large part to the capabilities of GIS" (MNCPPC 1998b, 11). It will be further reviewed later in this chapter.

Second, the county has also added to its GIS a countywide parcel coverage at a scale of $1'' = 200'$. GIS is now actively used to support master planning, environmental planning, and parks planning and is being integrated into other growth management functions (Schlee 1998).

Third, a consortium of county agencies has been formed to coordinate the development, management, and financing of the countywide GIS system (Schlee 1998). The consortium, called MC:MAPS, includes the executive branch, Park and Planning, the public and community schools, environmental protection, two incorporated jurisdictions, and other entities. Montgomery County is unique in the State of Maryland in that the State Department of Assessment and Taxation has turned over the maintenance of the tax maps to the county.

Annual Growth Policy

Montgomery County regulates approvals of development based on tests for adequacy of public facilities. The primary tests in recent years have been for transportation and public schools, inasmuch as other facilities, including water and sewerage facilities and police, fire, and health services have been found to be adequate (in the early years of this regulation, sewerage was the major constraint).

Each year, growth capacity ceilings are recommended for 28 Policy Areas, based on transportation capacity (Figure 4.5); see the draft *Fiscal Year 1999 Annual Growth Policy (AGP) Ceiling Element* (MNCPPC 1997). Policy Areas were initially created from the county's urban and suburban planning areas in order to implement the Adequate Public Facilities Ordinance during the 1970s. Since then, new Policy Areas have been established: Some of the original areas have been subdivided, and new Metro-Station Policy Areas have been delineated where additional development is encouraged adjacent to stops on the Metro rail transit line (Orlin 1998).

Policy Areas are not specified for the rural areas of the county, where urban development is not encouraged. In addition, the county's Annual Growth Policy does not apply to two incorporated municipalities, Gaithersburg and Rockville, which together include about 11 percent of the county's population. However, the county does monitor development, and calculate

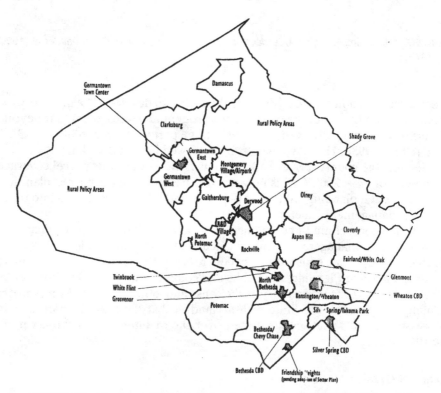

FIGURE 4.5   Montgomery County Policy Areas (*Source:* MNCPPC. 1997, p. 21)

and report staging ceilings for housing units and employment, within these municipalities in order to account for the impact of development and transportation service levels in these areas on the unincorporated areas of the county (MNCPPC 1997, 3). Notably, both Gaithersburg and Rockville were considered "moratorium areas" in 1997 because the amount of approved development in those towns creates more traffic congestion than the AGP's standard; however, the county does not have jurisdiction to enact a moratorium in these municipalities (Orlin 1998).

As mentioned, the main constraint on development is the transportation test. Transportation staging ceilings determine the total or gross amount of development (existing, approved, and to be approved) that can be handled by the transportation network without exceeding standards for road congestion. The net capacity, after subtracting existing and pipeline (approved but unbuilt) development, determines the amount of new residential (housing units) and employment (jobs) development that can be approved. As new transportation facilities (road or transit) are built, the ceilings are recalculated.

Two types of transportation tests are applied. First, the Policy Area Transportation Review determines staging ceilings, establishing whether there is sufficient transportation capacity in a Policy Area to accommodate additional preliminary plan approvals. Greater traffic congestion is permitted in areas with greater transit service and usage. This test considers the traffic impacts of existing and approved but unbuilt new development (the pipeline). Based on this review, the Montgomery County Council each year establishes job and housing staging ceilings for each Policy Area.

Second, the Local Area Transportation Review determines whether, for example, a proposed preliminary subdivision plan will cause unacceptable local traffic congestion at nearby critical intersections. If this will occur, the applicant may propose intersection improvements or provide trip reduction measures to offset traffic impacts.

Two types of staging ceiling flexibility are provided. "Full-cost developer participation" allows a preliminary plan to be approved in areas with insufficient staging ceiling capacity. In this case, the applicant agrees to pay for a road, transit, paratransit, or ride-sharing program that will add as much capacity to the system as the development will require. "Development district participation" allows private developers to request the County Council to create development districts to fund needed public facilities and to finance infrastructure over a longer term. Public–private partnerships to build infrastructure are permitted.

As of September 1997, countywide net staging ceiling allowed room for some 35,000 housing units and 32,000 jobs. Of the 28 Policy Areas, 6 were at capacity for housing and 5 for jobs, effectively triggering development moratoria in these Policy Areas (MNCPPC 1997).

School capacity is tested separately for the county's 21 high school clusters (Figure 4.6) and for elementary–middle school clusters. School capacity is considered adequate if forecast enrollment does not exceed 110 percent of

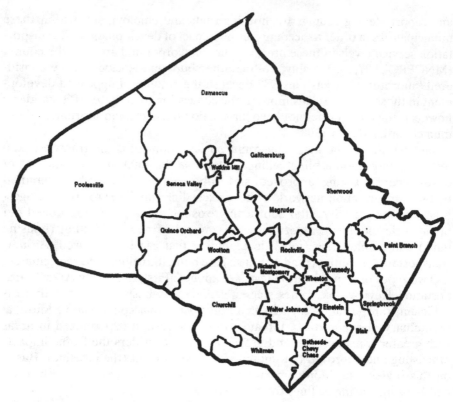

FIGURE 4.6   Montgomery County public high school clusters (*Source:* MNCPPC. 1993, p. 17)

the funded program capacity. If a school cluster or any of the clusters adjacent to it do not have sufficient capacity, then a preliminary school plan for that cluster cannot be approved in the next fiscal year. As of the 1999 report, all school clusters have adequate capacity at all three grade levels to support the 2002 forecast (MNCPPC 1997, 13). Because of a county priority for funding needed school capacity, no development moratoria have been declared owing to lack of school capacity (Orlin 1998).

Site Characteristics Inventory

The purpose of the *Site Characteristics Inventory* (MNCPPC 1998a) is to provide a definitive snapshot of the quantity of approved and developable nonresidential properties that could facilitate economic development and master plan build-out. The inventory serves to assess the status of economic development potential in the county, to evaluate the compatibility of existing land use patterns with economic development needs, and to assist the Department of Economic Development in economic recruitment.

The report presents a detailed evaluation of vacant and redevelopable nonresidential sites suitable for new commercial and office development.[4] These sites were initially identified in the 1996 report, *Economic Forces That Shape Montgomery County, Phase I,* which did not include environmental or other site constraints (MNCPPC 1998b). The final *Site Characteristics Inventory* screens vacant and redevelopable properties for environmentally non-buildable areas (i.e., wetlands, steep slopes, or stream buffers) and reviews the commercial projects in the pipeline (approved but not built), including the status of required transportation improvements for these projects. "Vacant" land is defined as undeveloped land zoned for employment use, but not in the commercial pipeline. "Redevelopable" sites have a land assessment for tax purposes greater than their building assessment, hence reflecting sites not developed to their potential. The inventory of redevelopable sites excluded the following: sites with a total floor area equal to or larger than that allowed by regulation, properties where demolition costs would effectively reduce the land value to below that of the building, and sites deemed not redevelopable because of location and surroundings.

Findings are presented for seven units called Community Based Geographic Areas, rather than for individual sites. The analysis finds that the county has ample types and locations of vacant and redevelopable land (39.7 million square feet) and pipeline projects (35.2 million square feet) to support short- and long-term business growth. This potential for another 75 million square feet of commercial space compares to the 152 million square feet of presently developed commercial space in the county. Furthermore, transportation improvement requirements and environmental conditions are not a significant constraint (MNCPPC 1998a).[5]

The 1998 Approach

Montgomery County's land supply monitoring approach underpins its outstanding growth management system. We define growth management as the tools and techniques used in a conscious government program intended to influence both the primary characteristics (rate, amount, type, location, and quality) and the secondary impacts (environmental quality, fiscal efficiency,

---

[4] With the construction of a new subway in the county, there has been intense interest in redevelopment possibilities in station areas. Redevelopment in some of the older central business districts (CBDs) has been hampered by small parcel sizes in these areas. It is hoped that the results of the *Site Characteristics Inventory* will provide valuable information for efforts to facilitate land assembly and revitalization more generally in these areas.

[5] The next phase of the capacity analysis carried out for the *Site Characteristics Inventory* will explore how this inventory relates to the needs of employers as potential users of nonresidential lands. The end result will be the crafting of a set of strategic recommendations to better coordinate and conform site availability with economic objectives.

and social equity) of future development within a local jurisdiction (Godschalk et al. 1979).

In general, growth management programs may include a statement of growth policy, a development plan, and various implementation tools— regulations, administrative devices, taxation schemes, public investment programs, and land acquisition techniques. A common feature of such programs is a desire to actively guide growth rather than simply to react to development proposals. An accurate and up-to-date inventory of land supply is essential for growth management programs to be effective and to avoid negative side effects, such as inflated land and housing costs, spillovers into adjacent jurisdictions, and the like.

As described earlier, the Montgomery County growth management approach includes a number of tools. Table 4.1 identifies these tools and the characteristics of growth they influence. The overall approach links the various tools as a coordinated package for guiding development to serve the public health, safety, and welfare.

A number of land supply monitoring tools also support the growth management approach. Although these tools are not organized as a closely coordinated system, they are linked within the planning organization. The tools, their database technologies, and the responsible organizations are described in Table 4.2. Many of the land supply monitoring tools that were operating in 1985 were still in place in 1998, with most of the responsibilities allocated between two divisions, Research and Development Review. As of 1998, much of the analysis and database management continued to be done with the original in-house software created in the 1980s, which was reviewed and up-

TABLE 4.1 Montgomery County Growth Management Tools

Tools	Growth Characteristics Affected	Comments
General plan	Type, location, amount of development	Urban goal for relationships between land uses, intensities, and infrastructure
Zoning	Type, location, amount, quality of development	Site arrangements, including height, bulk, setbacks
Subdivision regulations	Location, amount of development	Land conversion
Adequate Public Facilities Ordinance	Rate (timing) of development relative to infrastructure availability (especially transportation)	Short-term market influence; related to capital improvements program
Transfer of Development Rights	Location, amount of development	Equity provision; legal defensibility for open space down-zoning; incentive for agricultural preservation

TABLE 4.2 Montgomery County Land Supply Monitoring Tools

Function	Current Database Technology	Responsible Organization
Pipeline monitoring (approved, yet unbuilt projects)	Lotus spreadsheet	Research Division
Annual Growth Policy Ceilings	Lotus spreadsheet	Research Division
Commercial site identification	Lotus spreadsheet, ArcView GIS	Research Division
Subdivision approval	Development Management Information System (DIMS)	Development Review Division
Parcel file	ArcView GIS	Research Division
Transfer of Development Rights (TDRs)		Development Review Division
Building permit approval	Permit database system	Department of Permitting Services
GIS system and database development	Depends on user's system, but includes ArcView, ArcInfo, and others	MC: MAPS (Montgomery County: Mapping Automation and Preparation System)

dated to make it more user-friendly. Many of the operations were still being carried out on spreadsheets. The county had developed some GIS applications, and monitoring tools were being linked within a coordinated GIS system. Work was in progress to relate the parcel coverage and database file with the Adequate Public Facilities Ordinance and to reengineer these programs.

The Godschalk et al. (1986) study identified four stages of the life cycle of an Automated Land Supply Information System:

1. *Planning,* development of a course of action and project control guidelines, consideration of resource and time constraints
2. *Analysis and design,* coordinating work flow and user objectives and constraints, data requirements and user needs, system design
3. *Implementation,* establishing system procedures, dealing with transition problems, orienting users
4. *Operation,* periodically reviewing design efficiency and effectiveness, ensuring proper system function

Montgomery County is now in the fourth stage of its system's life cycle: integrating new GIS capabilities with the original software programs and redesigning those programs to be more efficient and effective under present conditions and to meet current needs.

Transferability of the Montgomery County Approach

In theory, none of the elements of the Montgomery County approach should be out of reach of other jurisdictions desiring to replicate it. In practice, however, replicating the complete growth management approach would be difficult for local governments in states without growth management laws and policies, and for localities without the necessary commitment from their political leaders to adopt a growth management program. It may also be difficult to replicate in politically fragmented counties with a number of incorporated municipalities. Montgomery County has the advantage of having to deal separately with only two sizeable municipalities, Gaithersburg and Rockville. Nonetheless, the general principle of enacting a general plan with a similar array of implementation tools should be transferable.

It should also be feasible to replicate the county's land supply monitoring approach. As many local governments are computerizing their assessors' databases and implementing complementary GIS programs, the challenge is less one of technology than one of succeeding at institutional innovation—see Nedovic-Budic and Godschalk (1994, 1996), for analyses of implementation influences and a case study of the adoption of GIS within a county government. It is especially important that government departments coordinate their functions, such as growth management and pipeline tracking, and share data in digital formats, such as data on subdivision approvals and adequate public facilities thresholds. In this regard, the concepts pioneered by Montgomery County should be transferable.

Issues Raised by the Montgomery County Experience

Issues raised by the Montgomery County experience, which relate to both technology and methods of growth management, include the following: (1) the trade-offs between using in-house and commercial software, (2) the difficulty of changing a computerized land supply system once it has been initiated, (3) the effect of defining acceptable and relevant geographic subareas for growth management, (4) the integration of public- and private-sector land supply databases, and (5) the potential impacts of growth management on land prices and spillovers to adjacent jurisdictions.

1. Montgomery County staff wrote the original programs needed for its land supply monitoring system. The county is in the process of reviewing, updating, and integrating its original in-house land supply software, of tying its programs into the GIS database, and of reengineering its system to use a client server. For localities starting new systems, it makes sense to purchase commercial software for these purposes, inasmuch as that software is supported by its issuing company and most localities will not have the necessary staff resources to create original software. The use of commercial software products should improve affordability and cost-effectiveness. However, all

local governments entering the computer mapping and records era should expect to face continuing transitions over the long term, as the development of new technology will continue.

2. The Montgomery County experience points to the relative permanence of a land supply monitoring system once in place and applied. Some 13 years after the initial study of the county's system, the county is still using versions of most of the same computer programs written initially. Although GIS was used in the recent *Site Characteristics Inventory,* spreadsheets and hand operations continue as primary land supply monitoring tools. However, changes are occurring. As of 1998, the county system was going to a client server, and the development review system was to be linked to GIS. The older computer systems were expected to be replaced by an HP 3000 and to be reengineered to use GIS with an Oracle database. A needs assessment was being conducted to write specifications for a new system to be designed in response to a Request for Proposals. This work was expected to take about two years (Schlee 1998).

3. As subareas, such as the Policy Areas, are defined to calculate development capacity, do these become mini de facto growth boundaries? Montgomery County officials understand the growth boundary to be not the delineation of land within Policy Areas, but the line between rural and suburban areas (the line between the wedges and the corridors) (Orlin 1998). However, this line is not referred to by the county as a growth boundary. Moreover, its application is different from that of Portland, Oregon, where the boundary is an elastic line that can expand outward to accommodate future growth. In contrast, the line between rural and suburban areas in Montgomery County is not considered movable, but a permanent dividing line based on an in-depth analysis of land use and suitability for agriculture and urban development. It is possible, although not likely, that Montgomery County's corridors could reach build-out. As the 1993 plan states, land use and zoning are adjusted in the context of the needs and features of individual planning areas. Reaching zoning ceilings includes the redevelopment of all existing properties to their maximum, plus maximum development of all vacant land despite site constraints (MNCPPC 1993, 5, 6).

4. Montgomery County has not integrated its land supply database with private-sector databases, such as the Multiple Listing System. However, the county does subscribe to some private listings of commercial and retail real estate and has employed a consultant to evaluate the commercial and office sites identified in its *Site Characteristics Inventory.* In the coming age of web-based information systems, users will likely expect to be able to link public- and private-sector land supply databases. This should be possible, especially if private-sector database managers employ parcel identification numbers (PINs) (or tax account numbers in Montgomery County) matching those used by public-sector database managers.

5. Economic theory asserts that, assuming ongoing demand, growth restrictions will drive up housing prices. It also asserts that land supply limits will

transfer development demand into adjacent jurisdictions not affected by growth mandates, causing a spillover effect. A study of housing prices in 17 planning areas in Montgomery County between 1982 and 1987 attempted to measure the effects of land use constraints. Pollakowski and Wachter (1990) found that land use regulations did raise prices for housing and developed land within the county, and that spillovers affected adjacent counties. However, they acknowledged that the effects of positive amenities on quality of life resulting from managed growth might also raise land and housing prices through an increase in demand. They concluded that the policy implications of their findings were not clear-cut and stated that net benefits from managed growth may still be positive, despite impacts on land and housing prices and spillovers.

MARYLAND'S SMART GROWTH LEGISLATION AND MDPROPERTY VIEW

Recently, the State of Maryland enacted Smart Growth legislation that requires local governments to direct state funding to special targeted areas to reduce urban sprawl.[6] It also provided a statewide parcel-based GIS system to local governments to assist in this process. Although Montgomery County is a partner with the state in this effort, the State Smart Growth initiative is expected to have a greater impact on Maryland's less developed counties than on the urbanized counties (Montgomery, Prince George's, and Howard), which already have antisprawl plans and programs as well as digital parcel databases.

Maryland's statewide parcel-level GIS information system, MdProperty View, was developed by the Maryland Office of Planning in conjunction with the Maryland State Department of Assessments and Taxation. MdProperty View allows users to access a jurisdiction's property maps and parcel attribute information with a standard Pentium PC using ESRI ArcView software. It stores parcels as a point coverage, with an overlay of parcel boundaries derived from scanned assessors' maps. Counties received the system in the fall of 1997 as a tool to assist in the designation of "Priority Funding Areas," which are deemed eligible for state grants for infrastructure under the Smart Growth Areas Act of 1997.

---

[6] The State of Maryland is one of the most urbanized states in the country. Its relatively small size (12,297 square miles) and high overall population density (512 persons per square mile) have helped to focus state-level attention on issues related to land use and land supply in urban and urbanizing areas. At the same time, significant parts of the state are still devoted to productive agriculture, a traditional land use that is at risk of displacement by rapidly spreading urban development. At present, more than 85 percent of the state's five million residents live in the urban corridor comprising Baltimore, the Washington, D.C., suburbs, and areas in between.

Smart Growth and Priority Funding Areas

As advocated by Governor Parris Glendening in his January 1997 State of the State address, the Smart Growth Areas Act of 1997 (Chapter 759 of the Laws of Maryland) is designed to attack the problem of suburban sprawl and protect cities and rural areas for tomorrow. It builds on the Economic Growth, Resource Protection, and Planning Act of 1992, which established new visions for the growth and development of Maryland's future. The first three visions adopted in 1992 guide the location of growth through (1) the general concentration of development in suitable areas, (2) the protection of sensitive areas, and (3) in rural areas, directing growth to existing population centers along with protection of resource areas.

The Smart Growth Act requires the state to target funding for growth-related projects to Priority Funding Areas, or PFAs (Maryland Office of Planning 1997a). The Act pre-identifies certain types of PFAs, including existing municipalities, land inside the Baltimore and Capitol Beltways, and enterprise zones. Counties then designate their local PFAs in their plans, based on land use, water and sewer service, and residential density criteria. Local governments are advised to have a formal PFA mapping and certification program within their planning departments, to establish rules and procedures, and to use their comprehensive plans to guide PFA mapping. Types of areas eligible for PFA designation include the following:

- Areas zoned for industrial and other employment uses
- Existing communities with sewer service
- Existing communities with water service only
- Areas beyond the periphery of developed portions of existing communities
- Areas other than existing communities (within locally designated growth areas)
- Rural villages (designated in county comprehensive plans)

Counties also can certify PFAs based on their own analyses of the supply of land needed to satisfy demand for development at densities consistent with their comprehensive plans (Maryland Office of Planning 1997a). Land capacity is estimated by using the holding capacity of vacant land based on comprehensive plan and zoning ordinance designations. The following are recommended steps to establish capacity:

1. Multiply maximum allowed residential density for each zoning category by the vacant acreage in that category by average household size to arrive at total population and dwelling units that can be accommodated; sum across all zoning categories.

2. Determine net values by subtracting approximately 20 percent (depending on local experience) of the developed land for streets, sidewalks, and utilities.
3. Factor in "under-build" to account for development that often occurs at lower than allowed densities (experience with each type of zoning district determines specific discounts).
4. Consider the potential of redevelopment to add capacity.
5. Consider the potential of infill to add capacity.
6. Project the population for the target date and determine whether the land available for development meets the need generated for residential and employment uses.

The Act focuses on residential land needs, with nonresidential land uses playing a smaller part. For determining commercial land needs, the guidelines recommend using an average number of commercial acres needed per 1,000 additional population.

To facilitate production of standardized maps for use by state agencies, counties are asked to use the MdProperty View system where possible. They are to digitize their PFAs directly into a new ArcView layer with the 1 to 2,000 scale State Highways Administration road layer as a background. If paper maps are produced, the Office of Planning will convert them to digital files.

The Smart Growth program should work well, given Montgomery County's long history of integrating public facility funding with the location and timing of new development. Priority Funding Areas should easily be identified in accordance with the adopted General Plan and Policy Area plans (Moritz 1998).

MdProperty View

MdProperty View is a statewide parcel-point GIS made available to all counties in CD-ROM format to support their compliance with the Smart Growth legislation, especially for designating Priority Funding Areas. The centralization of property assessment in a single state agency facilitated the compilation of this statewide GIS parcel database.

MdProperty View is a visually accessible database that allows users to access information by looking at a map on a computer screen, zooming in on an area of interest, and clicking on a point within the property parcel to display the parcel's database record (Maryland Office of Planning 1996). Users can also query the database to display properties with certain characteristics, so that, for instance, a map can be created of all parcels on a particular tax map that are vacant, commercially zoned, and at least 3 acres in size. Maps can be generated in various sizes, including $8\frac{1}{2}$ by 11 inches and 11 by 17 inches, at reduced, 100 percent, or enlarged scales, with the use of PC-compatible printers.

Starting in 1993, the state's Office of Planning in conjunction with the State Department of Assessments and Taxation, computerized 2,800 property maps, linking each digitized parcel point to its corresponding database record. Parcels were aligned by means of a GIS overlay with the State Highway Administration's digital road maps and with satellite imagery, creating the first computerized infrastructure base map for the state. The database is maintained by the Planning Data Services Division of the State Office of Planning.

The property database contains more than 60 data items. Among these are the owner's name and address, premise address, deed references, parcel number, land use, land value, structure size, improvement value, parcel size, year built, and trade date.

The primary impact of MdProperty View should be felt in the less developed counties of the state, which have not adopted plans to curb sprawl and preserve natural resources and have not been monitoring their land supply through computerized parcel databases. MdProperty View is also expected to increase access by developers and the business community to computerized parcel data. Montgomery County now sells the state parcel data in AutoCAD format, but is planning to sell ArcInfo coverages as well.

Transferability of Maryland's Smart Growth and MdProperty View

Maryland is unique in that its property assessment function is centralized at the state level, enabling the state to provide a statewide property database.[7] Its Office of Planning conducted a statewide GIS capacity analysis in the early 1990s, summarized in *The Potential for New Residential Development in Maryland: An Analysis of Residential Zoning Patterns* (Maryland Office of Planning 1992). The completion of the statewide parcel coverage should help with any future updates of this capacity study.

Most property assessment in other states is done by local government offices, however, and most states do not provide either a statewide assessment database or GIS software comparable to MdProperty View. It will be interesting to monitor the future development of such capabilities in other states, especially the current growth managing states and states that decide to emulate Maryland's Smart Growth antisprawl policy.

CONCLUSIONS

The way that land supply monitoring can be used to implement growth management depends on the type of growth management tools in use. There is a

---

[7] Other states that maintain statewide parcel information systems include Florida, whose Department of Revenue aggregates county assessment data, and South Carolina, whose University of South Carolina Humanities and Social Science Computing Laboratory maintains property and infrastructure files for all localities in the state for use in industrial recruiting and other functions.

major difference between growth management programs based on expandable Urban Growth Boundaries (UGBs), such as those used in Oregon and Washington, and programs, such as Montgomery County's, based on fixed greenbelts. The latter employ Adequate Public Facility Ordinances (APFOs) to stage the growth within designated urban areas and, in coordination with Transfers of Development Rights (TDRs), to preserve land and agriculture within designated rural areas. With a UGB approach, land supply monitoring provides a means to determine whether there is a need to move the urban growth limit further out in order to accommodate the projected population growth, or to revise permitted development patterns. With an APFO approach, land supply monitoring provides a means to stage the timing and location of growth within the urban area, by permitting growth only where adequate public infrastructure capacity exists.

There is also a difference between state growth management approaches that attack sprawl through requirements for concurrency of infrastructure and development at the local level, as in Florida (DeGrove 1992), and those that selectively target state funding for local growth-related projects, as in Maryland with Priority Funding Areas (PFAs). On the surface, the Maryland approach appears to add another arrow—control over the location of state project funding—to the local growth management quiver. However, the listing of types of areas eligible for PFA designation seems quite broad. The Maryland Smart Growth Act allows local governments considerable latitude and is not as rigorous as an Adequate Public Facilities Ordinance or a local concurrency requirement. It remains to be seen whether PFAs will be effective in curbing sprawl. The outcome may depend on the success of the Maryland Office of Planning in persuading counties to limit their PFAs to the size needed to accommodate projected growth within compact urban areas.

Land supply monitoring, which seemed somewhat unique in 1985 when only two dozen public systems could be identified (Godschalk et al. 1986), now appears to be gaining momentum as a standard operation of local governments engaged in growth management. Montgomery County continues to demonstrate the power of a pioneering approach that utilizes land supply monitoring to support its growth management tools. Land supply monitoring has also been given a major push at the state level by the Maryland Smart Growth Act and MdProperty View, which demonstrate the logic and power of a state's providing authority and tools to its localities in order to manage growth and development in the public interest.

REFERENCES

De Grove, John M. 1992. *The new frontier for land policy: Planning and growth management in the states.* Cambridge, Mass.: Lincoln Institute of Land Policy.

Godschalk, David R., Scott A. Bollens, John S. Hekman, and Mike E. Miles. 1986. *Land supply monitoring: A guide for improving public and private urban development decisions*. Boston: Oelgeschlager, Gunn & Hain, in association with the Lincoln Institute of Land Policy.

Godschalk, D. R., D. J. Brower, D. Herr, L. D. McBennett, and B. A. Vestal. 1979. *Constitutional issues of growth management*. Rev. ed. Chicago: Planners Press.

Maryland National Capital Park & Planning Commission (MNCPPC). 1980. *Functional master plan for the preservation of agriculture and rural open space in Montgomery County*. Silver Spring, Md.: MNCPPC.

———. 1993. *General Plan refinement of the goals and objectives for Montgomery County*. Approved and adopted. Silver Spring, Md.: MNCPPC.

———. 1995. *Annual growth policy*. Preliminary draft. Silver Spring, Md.: MNCPPC.

———. 1997. *Annual growth policy ceiling element. Fiscal year 1999*. Silver Spring, Md.: MNCPPC.

———. 1998a. *Report: Site characteristics inventory*. Silver Spring, Md.: MNCPPC.

———. 1998b. *Economic forces that shape Montgomery County: Annual update*. Silver Spring, Md.: MNCPPC.

Maryland Office of Planning. 1992. *The potential for new residential development in Maryland: An analysis of residential zoning patterns*. Managing Maryland's Growth: Issue Papers. <http://209.116.30.10/mmg/mmg.htm> (January 13, 1998).

———. 1996. *Database: Maryland State data center newsletter*. April.

———. 1997a. *Managing Maryland's growth: Models and guidelines. Smart Growth: Designating priority funding areas*. Baltimore: Maryland Office of Planning.

———. 1997b. *Maryland Property View*. <http://www.op.state.md.us/data/mdview1.htm> (January 2, 1998).

———. 1997c. *Smart Growth and neighborhood conservation in Maryland*. <http://www.op.state.md.us/smartgrowth/index.html> (October 8, 1997).

———. 1998. *Smart Growth options for Maryland's tributary strategies*. Baltimore: Maryland Office of Planning.

Moritz, C. (Growth Policy Coordinator, Montgomery County Department of Park and Planning). 1998. Telephone interview by author, May 15.

Nedovic-Budic, Zorica D., and David R. Godschalk. 1994. Implementation and management effectiveness in adoption of GIS technology in local governments. *Computers, Environment and Urban Systems* 5: 285–304.

———. 1996. Human factors in adoption of geographic information systems: A local government case study. *Public Administration Review* 6: 554–567.

Orlin, G. (Deputy Director, Montgomery County Council Staff). 1998. Telephone interview by author, June 26.

Pizor, Peter J. 1986. Making TDR work. *Journal of the American Planning Association* 2: 203–211.

Pollakowski, Henry O., and Susan M. Wachter. 1990. The effects of land-use constraints on housing prices. *Land Economics* 3: 315–324.

Schlee, J. (GIS Coordinator, Montgomery County Department of Park and Planning). 1998. Telephone interview by author, May 18.

COMMENTARY: CONTRASTING ISSUES FACED BY MONTGOMERY COUNTY AND PORTLAND

Lewis D. Hopkins

Database Implications of the Montgomery County, Maryland, Approach

Montgomery County's well-developed system of growth management relies on a set of techniques and tools, each suited to a particular aspect of managing urban development. The county has been using its current approach for almost 20 years and is constantly refining the approach and the analytical tools for carrying it out. Separate computing tools have been developed by individual departments and divisions for specific tasks.

The county is now working on the design and development of a new, integrated computer system to support these tasks. This provides an opportunity to consider briefly three questions. What, in very general terms, data and system design requirements support the Montgomery County system of growth management? How might these requirements differ from specifications to support growth management in Portland, Oregon, or elsewhere in the Northwest, where the focus is on Urban Growth Boundaries? To what extent can these requirements be met by traditional GIS and, in particular, how can parcel-based GIS support the county's growth management tools?

A standard approach to information system design involves four steps that address in turn (1) individual user views, (2) aggregate of user views, (3) logical data models, and (4) the requirements of database implementation. The focus here is on user views and some implications for logical data models.

For the purposes of this discussion, the four implementation tools that make up the current growth management system—the Transfer of Development Rights Program, the Zoning Ordinance, the Adequate Public Facilities Ordinance, and the Annual Growth Policy—represent individual user views. In designing an actual computer system, these user views would be disaggregated further into specific tasks, implying additional user views for such tasks as issuing subdivision and building permits, managing capital budgets, and providing infrastructure. Each of these views requires certain data entities. An entity is an object that has attributes. A tax-lot parcel is an entity, and its size is an attribute. Polygons of homogeneous characteristics or dwelling units are also entities. A database is designed by identifying the entities and attributes needed for each of the analyses to be supported, as follows:

The Transfer of Development Rights (TDR) Program permanently reallocates development rights from "wedges" of largely undeveloped land to "corridors" designated for urban development. TDR establishes regulations that support a market in which owners in the corridors have an incentive to purchase rights from owners in the wedges. (The owners in the corridors can thus develop at higher densities than would otherwise be allowed, and the owners in the wedges must develop at lower densities than could be enforced through the police power of zoning alone.) The TDR Program will require data entities for land parcels and for rights. Rights may exist separately from parcels if they are held in rights banks in order to ensure adequate prices.

The Zoning Ordinance sets the locations and densities for particular uses. Zoning applies to large areas but not necessarily to entire parcels, especially for large parcels awaiting development and when "overlay" zones such as floodplain regulations are considered. Thus, keeping track of zoned land will require spatial entities based on homogeneous zoning.

The Adequate Public Facilities Ordinance (APFO) sets timing by constraining development in areas not yet supplied with or programmed for infrastructure. The county's APFO focuses on "Policy Areas." Apparently, attribution of infrastructure-carrying capacity to particular parcels for which subdivision approval is sought is made on a first-come-first-served basis. Thus, until a subdivision application occurs, the primary entity of interest is the Policy Area. Levinson (1997) points out the difficulties of ascribing carrying capacity to Policy Areas, especially in the case of transportation capacity. For any given transportation network, there are many configurations of land uses that can be supported at a given level of service. Discovering land use patterns that are members of the feasible set, however, is difficult and subject to argument on methods and data. Just think of simple cases. An office building with 500 jobs could be located in any one of several different locations along a corridor running from the central city outward and use the same amount of services along that corridor. To which location should that level of service be attributed? Additional criteria must be considered, at least implicitly and preferably explicitly, in choosing how to attribute carrying capacity. In considering an entire transportation network, including intersection capacities and an entire land use pattern, the ambiguities are increased. It is clearly inappropriate to attribute carrying capacity to individual parcels. The transportation models used to estimate network capacity and to compute level of service consumed for any given proposal will be run for Transportation Analysis Zones, not parcels. Similar but less complicated estimates and computations apply for sewers, water, and schools. The primary entities for these analyses will be Policy Areas, Transportation Analysis Zones, sewer sheds, and school service areas.

The Annual Growth Policy links infrastructure plans and development constraints to capital improvements programming. The Annual Growth Policy and the capital improvements program will focus on projects and service networks. Data requirements will relate to computing budget feasibility and estimating the effects on infrastructure carrying capacity by policy region, or by some other mechanism of attributing network capacity to potential development proposals. This cursory consideration of user views suggests that several different types of entities, including parcels, will be needed to support these growth management tasks. System design and development should focus on meeting the needs of these various user views.

Database Implications of the Portland Metro Approach

The Montgomery County approach is distinctly different from the approach using Urban Growth Boundaries (UGBs), which is the focus of growth management in Oregon and, more recently, in the state of Washington. Simply stated, Oregon's UGB does the work of both Montgomery County's Transferable Development Rights Program and its Adequate Public Facilities Ordinance. The UGB confounds these functions and creates confusion, but the confusion solidifies a coalition of interest groups sufficient to support the continued imposition of growth boundaries. Planners have long recognized that we need not agree on ends in order to agree on means. Indeed, the growth boundary is seen by some as a means to the permanent or near permanent preservation of resource lands (for agriculture, forests, and recreation). To others, it is a means to delineate where infrastructure will be provided and development permitted in the near future.

Perhaps, in part, to help reconcile the conflict between pressures to treat the UGB as permanent and to treat it as regulating the timing of development, the Oregon legislation requires that the UGB include a 20-year supply of land. With the possible exceptions of sewage treatment plants, major sewer interceptors, and the location (but not the phased construction) of major transportation corridors, a 20-year land supply is too large an area in which concurrency can be achieved with the timing of infrastructure. Moreover, a UGB designed to include a 20-year supply of land does not specify a logic of urban form or size in the long run. Thus, rather than focusing on infrastructure capacity or regional form, as in Montgomery County, the most important user view in Oregon is determining whether there is a 20-year supply of land within the UGB. This question does not depend on transportation modeling or on infrastructure capacity, but on a definition of the supply of available land. This complex question is addressed in several of the chapters in this book. Hopkins and Knaap suggest that the task requires data entities of several different types.

A second user view in Oregon focuses on the monitoring of goal achievement. For example, are densities of development near light rail stations becoming higher than in other areas? Other questions, however, may be more pertinent. If the intent is to consume less land for urban uses in order to preserve resource land, then the question is whether aggregate consumption of land across all urban uses is lower, not whether parcel sizes in areas near transit stations are smaller. If the question is whether transit has gained a larger share of travel or whether total travel is lower, then the primary data entity will be Traffic Analysis Zones.

The most problematic task in implementing the Montgomery County approach is transportation modeling and interpretation of models in terms of capacity and annual growth policy. Assessing goal achievement for transit ridership or accessibility in Portland, Oregon, would also require estimating transportation behavior. Traditional GIS does not support transportation modeling or any kind of spatial interaction modeling above the level of very basic business applications. Traditional GIS also does not support dynamic data very well, and growth management is an inherently dynamic problem. Data structures specifically designed to support urban modeling and analysis by focusing on the modeled elements, rather than on maps, may be able to resolve the limitations of GIS (Hopkins 1999).

Conclusion

Godschalk's careful articulation of the functions of each element of Montgomery County's growth management system provides an excellent opportunity to specify requirements for a growth management support system. It also helps highlight some of the data requirements of the contrasting approach to growth management used in Oregon. The initial thoughts presented here about the design of such a system suggest that it should support several different spatial data entities in addition to tax lot parcels, and that it should support modeling tools that are not part of traditional GIS. It may be no accident that Montgomery County is at the forefront of growth management, but not at the forefront of implementing parcel-based GIS and traditional GIS tools.

References

Hopkins, Lewis D. 1999. Structure of a planning support system for urban development. *Environment and Planning B: Planning and Design* 3: 333–343.

Levinson, D. 1997. The limits to growth management: Development regulation in Montgomery County, Maryland. *Environment and Planning B: Planning and Design* 5: 689–707.

5

Central Puget Sound Region, Washington: Study of Industrial Land Supply and Demand

Lori Peckol and Miles Erickson

The central Puget Sound Region of Washington State is located between the Cascade and Olympic mountain ranges and is bisected by the saltwater inlets of Puget Sound. The region includes four counties, King, Kitsap, Pierce, and Snohomish, which altogether cover 6,300 square miles. Between 1960 and 1997 the region's population doubled, from 1.5 million to 3.1 million and is expected to reach 4.3 million by 2020.

After World War II the region's economic base, dependent on resource-oriented industries in the early twentieth century, evolved into manufacturing-dominated industries, including a robust aerospace sector. The economy is still strongly influenced by aerospace but has diversified significantly since the 1980s. Employment, which totaled 1.7 million in 1996, is forecast to reach 2.4 million by 2020. Twenty percent of projected job growth is in sectors tradition-

The authors wish to thank Norman Abbott, of the Puget Sound Regional Council, and Robert Filley, of the Center for Community Department and Real Estate, for their contribution to this chapter.

Monitoring Land Supply with Geographic Information Systems, edited by Anne Vernez Moudon and Michael Hubner ISBN 0 471371673 © 2000 John Wiley & Sons, Inc.

ally viewed as industrial: manufacturing, construction, wholesale trade, and transportation/communications/utilities.

Washington's Growth Management Act, enacted in 1990, requires the counties in this region and other fast-growing parts of the state to designate urban growth areas (UGAs) as part of an effort to reduce sprawl. In 1996, the UGA for the central Puget Sound Region totaled approximately 1,000 square miles.

The Puget Sound Regional Council (PSRC or Regional Council) is designated under federal law as the metropolitan planning organization (MPO), and under state law as the regional transportation planning organization (RTPO) for the region. State laws, including the Growth Management Act, and federal laws, including the 1991 Intermodal Surface Transportation Efficiency Act, require the central Puget Sound Region to have a regional growth management, economic, and transportation strategy. This strategy, VISION 2020, was adopted in 1991 and updated in 1995 to reflect local comprehensive plans completed pursuant to the Growth Management Act.

In early 1996, the Regional Council's Growth Management Policy Board[1] initiated a two-year study of industrial land supply and demand in the four-county region. The Policy Board regarded adequate industrial land supply as important to regional economic growth. In 1996 about a third of all jobs within the region were located on industrial land, and approximately 20 percent of anticipated new jobs between 1996 and 2020 are expected to be in manufacturing, wholesale, and other industrial sectors. As many of the most accessible and inexpensive locations in the region become developed, strategic use of the remaining industrial land supply is increasingly essential.

In fall 1996, the University of Washington's Center for Community Development and Real Estate (CCDRE) received funding to conduct a complementary study of demand for industrial land.[2] Because the two studies were intended to address several of the same questions, Regional Council and CCDRE staff worked closely together to integrate the work.

This chapter provides a summary and analysis of the Regional Council-CCDRE study of industrial land supply and demand (PSRC and CCDRE 1998).[3] Although not completely parcel-based, the methodology and scope of the study are significantly similar to other land supply and capacity analyses that have relied on parcel-level data. The first section briefly reviews the objectives and context for the industrial land supply and demand study. The second section summarizes the methodologies used in the study. The third section provides a brief summary of the supply and demand comparison

---

[1] The Growth Management Policy Board includes elected officials from member cities and counties, as well as representatives of business, labor, environmental, and civic groups.

[2] Funding agencies included the National Association of Industrial and Office Properties (NAIOP), the Society of Industrial Realtors (SIOR), the Port of Seattle, and King County.

[3] The complete study report can be acquired from the Puget Sound Regional Council's Information Center at 206-464-7532 or infoctr@psrc.org.

for the region. The fourth section discusses study limitations and potential enhancements.

BACKGROUND INFORMATION AND CONTEXT

Study Objectives

A series of five focus group meetings were held with local economic development experts and the Regional Council's Growth Management Policy Board members to provide input on issues to be addressed in the industrial lands study. Based on this input, the primary study objectives were established as follows:

1. *Seek agreement on a methodology for estimating supply and demand for industrial land that could be applied regionwide.* In the past, different methods were used by various jurisdictions and interests studying industrial lands. Data were difficult to aggregate at the regional level because of differing definitions and approaches for such items as redevelopment thresholds, building coverage ratios, space needs per employee, and the use (or not) of a "market factor."

2. *Achieve an in-depth understanding of long-term industrial land supply and demand in the region.* On the supply side, the project set out to identify designated industrial land, to estimate net supply (excluding land that could not be built on, such as wetlands and rights-of-way), and to identify the characteristics of major industrial areas, such as location, parcel sizes, price, and infrastructure. On the demand side, project objectives were to provide better data regarding the amount of building space used by employees in various industrial sectors and to convert forecasted employment growth to industrial land needs for the region. The study was intended not only to advance regional understanding, but also to provide information about the four-county industrial land context that might be useful to local jurisdictions.

3. *Based on data and information generated by the study, consider strategies that will ensure adequate industrial land supply in the future.* The project set out to make recommendations, within the context of adopted plans and the Growth Management Act, on strategies the region could use to best meet future demand for industrial lands.

The design of the study incorporated a number of limitations. First, because of the regional nature of the analysis, it did not compare supply and demand on a subarea basis or for specific industrial sectors. Second, it did not identify the supply of land currently for sale or land meeting the specific needs of a particular industry sector or user for infrastructure or transportation. Third, it did not address concerns that real estate brokers and buyers consider when negotiating a real estate transaction, such as soil conditions, parcel shape and configuration, opportunities for parcel assembly, cost, marketability, parking, or aesthetics. Finally, it was not designed as an ongoing monitoring study, although data collected in the study could be tracked over time.

Study Context

Several institutional and technical conditions existed in the central Puget Sound Region in late 1995 that contributed positively to this study. These conditions, discussed in the following paragraphs, may help to create a favorable environment elsewhere for undertaking a similar study.

Importance of Industrial Land Supply as a Regional Issue   In the initial discussions with focus group participants, it became clear that improving the understanding of supply and demand for industrial land throughout the region was of significant importance to a wide range of groups interested in land use and economic development. As a result, many people were willing to participate and contribute to the study.

Supply and demand for industrial land were also clearly seen as multicounty issues and appropriate for a regional agency to address. On the east side of the central Puget Sound Region, geography has historically contained east-west expansion between the Puget Sound and the Cascade Mountains. The primary industrial corridor in the region extends north-to-south across Snohomish, King, and Pierce Counties. On the west side of the Sound, Kitsap County's industrial development, such as the Puget Sound Naval Shipyards, has in part been defense-related. Elected officials and staff working on this project recognized that as the central Puget Sound Region continues to grow and competition from other uses increases, it becomes increasingly necessary to understand the dynamics of supply and demand for the entire regional market.

Recent Implementation of Growth Management   The Washington Growth Management Act requires the fastest-growing counties and cities to prepare comprehensive plans to protect open space and conserve resource lands, while managing growth. By early 1996 the majority of local jurisdictions had adopted comprehensive plans and development regulations. In the course of preparing these plans, many jurisdictions had inventoried existing land use and capital facilities. This work provided a valuable basis and reference for the industrial lands study. In addition, the Economic Development Councils for Snohomish and Kitsap Counties, and the Washington State Department of Community, Trade and Economic Development, had recently completed industrial land inventories that also provided useful data (Therrien & Price 1995; Rhine 1996).

Previous Work on Methodologies for Calculating Land Capacity   Much work has been done in the region on methods for calculating land capacity. For example, King County and its cities completed a countywide estimate of land capacity in the early 1990s. Following this estimate, the county formed a task force in 1995 to review and refine the methodology, which was applied between 1996 and 1997 (King County Land 1995; King County Office 1997). Snohomish County has completed several studies of employment land, the most recent of which is an analysis of employment land capacity revised in 1995 (Snohomish County Department of Planning and Community Development 1985-1994; Therrien & Price 1995). This and other work in the region provided a basis for the methodologies used in the industrial lands study. In addition, several members of a technical advisory committee for the industrial lands study had participated in capacity analysis studies and brought that knowledge and experience to the discussions.

Favorable Technical Conditions   Each of the counties and the major cities in the region use geographic information systems (GIS). The amount and types of data managed by these systems vary with both parcel- and non-parcel-based systems represented. Database development efforts for the study greatly benefited from the existence of local jurisdictions' GIS-produced land use and zoning maps. These data were integrated electronically to form a four-county database of the locations of designated industrial lands. In addition, some jurisdictions provided net supply estimates for the study, some of which were done using parcel-based GIS. For the industrial lands study, data on land supply were represented at the level of "concentrations" rather than individual parcels.

STUDY METHODOLOGIES

The methodologies for the supply and demand portions of the industrial land study are described in the following sections of this chapter. The "Supply" section discusses definitions and identification of industrial land, data collection and sources, and calculations of net industrial land supply. The "Demand" section discusses the model used to forecast future needs for industrial land in the region.[4]

Supply Methodology

The base for the industrial land supply estimate was urban land designated for industrial use in city and county comprehensive plans. These lands were further divided, for analysis purposes, into 69 major concentrations and 67

---

[4] Study methodologies and results were reviewed by a technical advisory committee (TAC). The TAC was created to draw together a regionally representative group of people who were knowledgeable about industrial land issues and to provide a mechanism for broad participation. It included representatives from commercial real estate firms, economic development councils, local government, and ports, civic groups, and the University of Washington.

small concentrations (see Figure 5.1 for map). The supply portion of the study included four primary steps.

Step 1: Define industrial land. Industrial land was defined as urban land designated in comprehensive plans for manufacturing, heavy and light industry, research and development, wholesale trade, warehousing, distribution, and business parks. Industrial land included in the study is located in 50 cities, towns, and counties and as parts of lands held three Native American tribes in the region.

Step 2: Identify industrial concentrations. The unit of analysis was concentrations of contiguous industrial land (developed or undeveloped) at least 25 acres in size. Smaller areas were excluded owing to the regional scale of the study. A total of 136 concentrations were identified through review of comprehensive plans. These areas were categorized by size as 69 major concentrations of 180 acres or more and 67 small concentrations of 25 to 179 acres.

Step 3: Collect data. For all concentrations, information was collected on acres of designated industrial land and net industrial land supply. For major concentrations, additional data were collected to determine the type of industrial area, land readiness for development, development activity, and potential for erosion of industrial land supply by nonindustrial development. Data included planning and zoning designations, industrial area categorizations, transportation and infrastructure service and adequacy, environmental hazards, parcel sizes and prices, and scale of sales or leasing activity.

Primary data sources included MetroScan assessor data, planning and public works departments of local jurisdictions, utility districts, and commercial brokers. Other recent industrial studies were also consulted. The data were collected between March 1996 and August 1997. Additional information regarding data collected for major concentrations is provided in Table 5.1.

Step 4: Calculate estimated net industrial land supply. The methodology used for estimating net industrial land supply, summarized below was based on King County's 1995 countywide land capacity methodology and was reviewed by the technical advisory committee (King County Land Capacity Task Force 1995). If jurisdictions had an up-to-date industrial land supply estimate that was generally consistent with this methodology, that estimate was used. Otherwise, Regional Council staff prepared the supply estimate using parcel data from 1996 and 1997.

- *Vacant and Redevelopable Parcels:* Parcel data for industrial concentrations were reviewed to identify vacant and potentially redevelopable parcels. Parcels were included regardless of whether they were potentially for sale or lease. Vacant parcels were defined as those with zero assessed

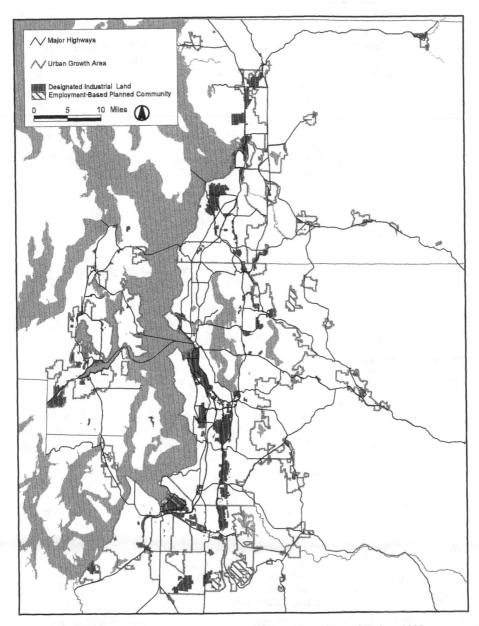

FIGURE 5.1 *Designated industrial land, Central Puget Sound region, 1996*

TABLE 5.1 Data Collected for Major Industrial Concentrations

Data	Description	Primary Source(s)
LOCATION AND JURISDICTION	Concentration name and jurisdiction within which it is located	Comprehensive plans, local jurisdictions
TOTAL LAND AREA (IN ACRES)	Includes developed and undeveloped areas, critical areas, street and utility rights-of-way, and other public purpose lands designated for industrial use	ArcView measurement or local jurisdictions
PLANNING/ZONING DESIGNATIONS	Designations, description, summary of permitted uses, restrictions on nonindustrial uses, and potential for conversion of industrially planned land to other designations (such as commercial or residential)	Local jurisdictions
PRIMARY OCCUPANTS	Names of major businesses or other users located at the concentration	Local jurisdictions
TYPE OF INDUSTRIAL AREA	Concentrations categorized by type as primarily manufacturing (light or heavy), warehouse, business park, or mixed (i.e., includes or expected to include multiple industrial uses)	Local jurisdictions

Transportation/Infrastructure

Data	Description	Primary Source(s)
ACCESS ROADS	Names of primary roads providing access to concentration	Puget Sound Regional Council data
DISTANCE TO NATIONAL HIGHWAY SYSTEM	Number of miles to nearest designated national highway, measured from a central point within the concentration	Puget Sound Regional Council data
RAIL	Whether the concentration is served or not served by rail	Puget Sound Regional Council data
WATER AND SANITARY SEWER	Service provider, brief description of service, and whether adequate for full development of the concentration, assuming typical industrial uses (if not adequate, description of proposed improvements and anticipated funding source)	Utility districts and local jurisdictions' public works and planning departments
STORM SEWER	Brief description of service and whether adequate for full development of the concentration (if not adequate, description of proposed improvements and anticipated funding source)	Local jurisdictions' public works and planning departments

TABLE 5.1 *(Continued)*

Data	Description	Primary Source(s)
TRANSPORTATION IMPROVEMENTS	Lists of projects (other than maintenance) from local and state six-year transportation improvement programs (TIPs) in vicinity of the concentration; with completion of these planned improvements, whether transportation service will be adequate for full development of the concentration, assuming typical industrial uses	Puget Sound Regional Council database of local transportation improvement programs, Washington State Department of Transportation six-year Transportation Improvement Program, local jurisdictions' public works and planning departments

Designated Critical Areas and Environmental Hazards

Data	Description	Primary Source(s)
WETLANDS, FLOOD HAZARDS, STEEP SLOPES	Acres of land unavailable for development because of designated wetlands, flood hazard areas, or steep slopes	Local jurisdictions
OTHER CRITICAL AREAS	Description of other critical areas at the concentration that affect land supply	Local jurisdictions
IDENTIFIED CONTAMINATED AREAS	Whether any part of the industrial concentration is on federal or state lists of contaminated sites; cleanup history as available	Environmental Protection Agency's National Priorities Sites List, Washington State Department of Ecology's Confirmed and Suspected Contaminated Sites List

Industrial Land Supply

Data	Description	Primary Source(s)
ESTIMATED NET INDUSTRIAL LAND SUPPLY	Estimated net supply expressed in acres of vacant and potentially redevelopable land and, as available, square feet of buildable floor area and additional employment capacity	MetroScan assessor data, local jurisdiction, and, in some cases, information provided by owners or developers; recent industrial land inventories conducted by the Snohomish County Economic Development Council (1995), Kitsap County Economic Development Council (1994), and Washington State Department of Community, Trade and Economic Development (1996) also consulted

TABLE 5.1   (Continued)

Data	Description	Primary Source(s)
PARCEL SIZES	Total number of vacant and rede-velopable parcels, largest parcel, plus number and area in the following categories: less than 2 acres, 2 to 9 acres, 10 to 19 acres, 20 or more acres	MetroScan assessor data, local jurisdictions
SCALE OF ACTIVITY	Amount of sales, leasing or development permit activity, expressed on a scale	Brokers
LAND ON THE MARKET	Estimate of amount of land on the market for sale or lease	Brokers
LAND PRICE	Typical price of vacant and improved land	Brokers

improvement value. Potentially redevelopable parcels were those with assessed improvement value less than 25 percent of total assessed value. Parcels identified as vacant or potentially redevelopable were further screened for use to subtract parks or common open space, utility or railroad use, street rights-of-way, and similar exceptions. The supply subtotal for redevelopable parcels was discounted by 10 percent to reflect the uncertainty associated with economically feasible redevelopment of these parcels.

- *Critical Areas:* Land unavailable for development because of designated wetland, flood area, steep slope, or other critical areas was subtracted either on a parcel-specific basis or at the concentration summary level based on jurisdictions' estimates or discount standards.

- *Future Street Rights-of-Way:* A discount for future street rights-of-way needs was subtracted at the concentration summary level. The discount was 10 percent for undeveloped areas, 5 percent for partially developed areas, and none for developed areas.

The data were mapped by concentration, with the use of ArcView in a series of thematic maps showing designated land, net acres of supply, 10-acre or larger vacant or redevelopable parcels, transportation accessibility, water and sanitary sewer service adequacy, and level of current and projected development activity. The data were also linked to other coverages, including 1995 employment data from the Washington State Employment Securities Department.

Demand Methodology

The demand methodology, which converts a projection of employment growth into an estimate of land needs, employed the following formula:

$$\text{Employment growth} \times \text{building square feet per employee}$$
$$\div \text{ coverage ratio} = \text{land needed}$$

To arrive at an estimate of total land needs, the demand formula was applied separately to each of four industrial sectors: construction, manufacturing, transportation/communications/utilities, and wholesale trade. In each instance, the total number of new jobs between 1996 and 2020 was multiplied by the industrial sector's current ratio of square footage per employee. The resulting projection of building space needs was then divided into six building types. For each building type, space needs were divided by an appropriate building-to-land-coverage ratio to determine the amount of land needed.

The methodology to estimate the demand for industrial land assessed two possible future use patterns. The first assumed that designated industrial land would be used for job growth in only the four industrial sectors identified here, and that 100 percent of the employment growth in these sectors would occur on industrial lands. The second use pattern assumed that employment growth, both industrial and nonindustrial, would be distributed between industrial and commercial lands in a pattern similar to that currently observed. To estimate this distribution, the Regional Council conducted a geographic analysis of 1995 employment data from the Washington State Employment Securities Department. This analysis identified the subset of employers located on industrial land. It then applied the percentage of employment occurring on industrial land by major Standard Industrial Classification (SIC) sector to projected employment growth. This second demand scenario yielded lower levels of industrial land consumption than the first.

Employment Growth and Land Use Patterns   The employment figures used in the model were derived from forecasts of regional employment growth by sector through the year 2020, which were developed by the Puget Sound Regional Council (Conway, Dick & Associates 1997).

Building Square Footage per Employee   Because industrial employment densities vary according to the type of industrial activity, CCDRE conducted a survey to determine the average square footage per employee for each of the four aggregate industrial sectors (Figure 5.2).[5] The survey asked Puget Sound industrial land users and property managers to provide the following information on their buildings: total square footage, number of employees, SIC code, and building type. The final survey data set included more than 1,000 buildings occupied by more than 600 companies.

[5] The second application of the demand model included projected growth in industrial and nonindustrial employment sectors (Puget Sound Regional Council and University of Washington Center for Community Development and Real Estate 1998, 6). For this application, the square-feet-per-employee ratio for the retail sector was derived from retail uses in CCDRE's survey. For the other nonindustrial sectors, the square footage per employee and building to land-coverage ratios used in this application are estimates based on review of other studies.

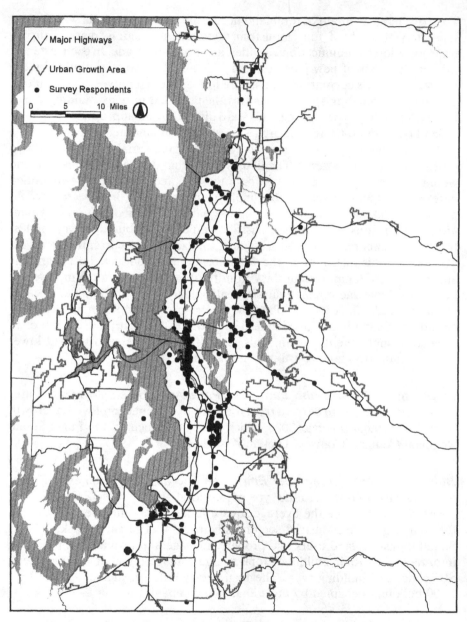

FIGURE 5.2 Location of CCDRE survey respondents

To determine the average square footage per employee for each industrial sector, the following formula was used:

Total square footage ÷ total employment = square footage per employee

Multiplying each sector's square footage per employee by the projected employment growth yielded an estimate of the total square footage *within buildings* that would be needed to accommodate employment growth between 1996 and 2020.

There are relatively few publications on available methodologies to determine ratios of industrial square feet per employee. However, CCDRE sought to check its survey results against other sources for such estimates and identified three sources that included square-feet-per-employee ratios for two of the building types highlighted in the Center's survey (Real Estate Economics 1984; Portland Metro 1990; Economic and Planning Systems 1994). As shown in Table 5.2, the Center's survey results generally concur with these sources.

An additional report by Cognetics, *America's Industrial Economy* (Birch et al. 1995), reviews approximate nationwide square-feet-per-employee ratios for several industries, including the four sectors targeted by CCDRE's survey. Figure 5.3 illustrates similarities and differences between the CCDRE figures and corresponding values from Cognetics. CCDRE survey results and corresponding Cognetics figures are remarkably similar for the wholesale and manufacturing sectors. For the construction and transportation/communications/utilities sectors, however, the results are significantly different, with CCDRE results indicating a need for nearly three times the square feet per employee suggested by Cognetics. There is no obvious explanation for these differences. One consideration lies in the large variation in square feet per employee reported by industries in the transportation/communications/utilities sector. CCDRE survey responses ranged from extremely low square-feet-per-employee ratios in the communications sector to extremely high ratios in the transportation sector. However, CCDRE and Cognetics had a similar proportion of responses from transportation industries that typically use the most space per employee. The difference between the Cognetics and CCDRE results is also large for the construction sector. This sector, however, accounts

TABLE 5.2   Comparison of Square Feet/Employee Study Results by Building Type

Building Type Surveys	Real Estate Economics (1984)	Portland Metro (1990)	Economic and Planning Systems (1994)	CCDRE Survey Data (1997)
MANUFACTURING	750	700	600 to 850	587
WAREHOUSE/ DISTRIBUTION	1,250	900	1,000 to 2,000	1,121

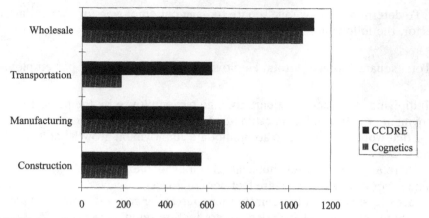

FIGURE 5.3   Square feet per employee by SIC category: Comparison of CCDRE and Cognetics study results

for less than 10 percent of the Puget Sound Region's projected employment growth through 2020.

Coverage Ratio   Industrial uses need land beyond the building footprint square footage to accommodate related uses such as setbacks, parking, truck maneuvering, storage, landscaping, and drainage. A building-to-land coverage ratio is the relationship between the footprint of a single-storey building to the buildable site. Coverage ratios followed this formula:

Building footprint ÷ lot size = building-to-land coverage ratio

The coverage ratios used in this study were established following a review of industry handbooks and discussions with local site planners, developers, and architects. Dividing the total square footage needed in each building type by the appropriate coverage ratio yielded the total land area needed, assuming single-storey construction (multiple-storey construction, which would require less land, is currently uncommon in Puget Sound industrial areas, with the exception of some business parks).

COMPARISON OF REGIONAL INDUSTRIAL LAND SUPPLY AND DEMAND

The study results indicated that through the year 2020, projected demand for industrial land in the central Puget Sound Region ranged between 5,600 and 7,100 acres. With the estimated net supply at 21,000 acres, supply exceeded demand by a factor of three. However, several important caveats were emphasized in the study report. First, the study found that a third of the supply was not currently served by infrastructure and adequate transportation. Second, the supply was distributed over a four-county area and was not predominantly located in the region's major industrial markets. Third, the evaluation did not compare supply and demand on a subarea basis and did not address questions of land adequacy for specific industrial sectors. Finally, the study emphasized the need to periodically monitor industrial land supply and demand.

CONCLUSION: LIMITATIONS AND POTENTIAL ENHANCEMENTS

The Washington State legislature amended the Growth Management Act in 1997 to require local jurisdictions to monitor land capacity and to determine whether sufficient suitable land exists to accommodate population and employment growth (see Appendix B). As cities and counties work to meet these requirements, the methodologies used will continue to evolve along with knowledge, data, and technology improvements. The following are some of the limitations in the industrial land study and potential enhancements that can serve as lessons for future industrial land supply and capacity monitoring activities.

Nature of Parcels Identified as Potentially Redevelopable

The study defined potentially redevelopable parcels as those with an improvement to total assessed value of .25 or less. Although defining "redevelopable" can be challenging for any kind of capacity study, the difficulties are exacerbated for lands used for manufacturing, construction, and other uses that involve outdoor storage. A portion of the parcels identified in this study as potentially redevelopable is used for such storage. Some of these "low value" areas are temporary, but others may be a critical part of existing business activities. Moreover, as prime industrial land is consumed, there may be increased pressure to make more efficient use of these areas.

Better attribute data are needed to indicate the actual use of parcels. For example, a parcel with low or zero improvement value may be used for storage or may be occupied by a portion of a building primarily located on another parcel, or it may indeed contain a business structure that underuses the property, given its value. Better information will help in assessing the potential redevelopability of parcels that may be further categorized to indicate a range of potential supply.

Publicly or Quasi-Publicly Owned Land

Vacant land zoned for industrial use but owned by cities, counties, states, or ports, is sometimes regarded as land that will not be available for private-sector use. Although jurisdictions may, in some cases, hold such land for future public use, they can and do also sell it for industrial development. If they do retain the land, it may be used for nonindustrial employment, such as provided in government facilities.

In this study most publicly owned land was included in the supply, with only a few jurisdictions choosing not to include such land. With certain exceptions (such as vacant land that is specifically committed for a public facility), publicly owned land should be included in industrial land supply studies, but should be differentiated to reflect the uncertainty of availability for private-sector use.

Developed Land

This study did not specifically track the amount of industrial land that is currently developed, but it can be roughly estimated. In future studies it would be useful to establish a base of how much industrial land is considered developed, as well as generally how it is developed (e.g., for industrial, commercial, residential, or public uses). To the extent that data are available, it would also be useful to estimate the historical rate of industrial land absorption and the type of use that consumed it.

Inventory of Industrial Land

As part of a second phase of work, the Regional Council and CCDRE conducted further review of the differences in the types and impacts of employment located on designated industrial versus nonindustrial land. One of the initial findings was that many jurisdictions include a commercial/industrial or heavy commercial zone that permits a number of light industrial uses, such as warehousing, wholesale, and light manufacturing. Clearly, distinguishing light industrial from commercial uses can be difficult, as, in many cases, the differences are inconsequential. For other industrial uses, however, service needs and impacts on neighbors differ sufficiently from those of commercial/office uses that they require separate consideration. Overall, these challenges suggest that including all land associated with employment in an employment land capacity analysis would provide more comprehensive and perhaps even more accurate results than achieved in this study.

Unit of Analysis for Characteristics of Land Supply

This study measured a number of characteristics of land supply, such as infrastructure adequacy and land price. Most of these characteristics were measured for concentrations of industrial land, several of which contained 500 or more acres. This relatively large unit of study simplified data collection and analysis. However, analysis at the parcel level or for smaller geographic areas would better capture variation in land characteristics.

Parcel data, as well as information concerning utilities, zoning, or critical areas, are currently not available digitally for the entire region. Each of the four counties in the region is in the process of establishing a countywide parcel coverage that will be integrated with assessor data. Establishing a parcel-level coverage and using GIS to combine it with other coverages (such as utilities, transportation, aerial images) would offer a number of advantages for analyzing the kind of parcel characteristics considered in this study. A parcel-level coverage would enable the kinds of data tracking and analysis listed below. At this level of resolution, database reliability is essential, and considerable time and resources would be necessary to conduct supply and capacity analyses of the following kinds:

- Analysis of the configuration or shape of parcels to screen for parcels that are not suitable for industrial development
- Spatial analysis of the location of vacant and redevelopable parcels to assess the potential to assemble parcels into tracts of 50 to 100 acres, able to accommodate large industrial developments
- Identification of uses on parcels that have zero improvement value but a use code that indicates the parcel to be at least partially developed
- Analysis of the potential capacity of partially developed parcels that meet the definition of "developed" but are only partly occupied by buildings, other site improvements, or critical areas
- A finer-grained analysis of environmental constraints and contaminated sites; analysis of the adequacy of utilities or transportation accessibility for small concentrations of a few parcels
- Analysis of industrial land supply in various zoning categories, such as heavy industrial, business park, or light industrial

Unit of Analysis for Survey of Industrial Land Users

The survey of industrial land users was needed to obtain basic information on building square footage per employee and building type distribution. Although this survey is believed to be accurate on a regional level, the sample size is not sufficient for subregional analysis and does not reflect subregional variations. In addition, the survey information gathered is static and does not account for changes in industrial development patterns over time.

Information similar to that collected by CCDRE gathered at the parcel level would provide more precise analyses. The data collected for each parcel could include the following:

- Total land area
- Square feet of building space
- Number of employees

- Standard Industrial Classification (SIC)
- Year developed

Through the use of parcel-based GIS, these data would create opportunities for new types of analysis, including (1) industrial densities calculated by subregion, (2) local concentrations of specific industrial uses, and (3) monitoring of ongoing changes or trends in land use patterns. This type of analysis could be applied to both industrial and nonindustrial land uses.

Monitoring Trends

As counties and cities establish the monitoring systems required by the 1997 Growth Management Act amendments, it would be useful to incorporate in these systems the means to track the following trends for future analysis:

- Acres of industrial land used annually in the region (by jurisdiction) and the type of use (industrial, commercial, residential, public).
- Amount of employment growth occurring on redeveloped versus vacant parcels.
- Amount of employment growth occurring within existing buildings and sites considered developed. This would involve both spatial analysis of employment growth and tracking the number of square feet added to buildings.
- Size of formerly vacant or redevelopable parcels used since the last evaluation, to track industry parcel use patterns.
- Changes in floor area ratios for industrial development, to track the amount of industrial development occurring in multistory buildings.
- Changes in building-to-land coverage ratios.

REFERENCES

Birch, David, Anne Haggerty, and William Parsons. 1995. *America's industrial economy.* Cambridge, Mass.: Cognetics, Inc.

Conway, Dick & Associates. 1997. *Regional economic and demographic database: Modeling and forecasting long-range forecasts for the Puget Sound Region.* Seattle. Prepared for the Puget Sound Regional Council. September.

Economic and Planning Systems. 1994. *Employment-based space projections model.* Berkeley, Cal.: September.

King County Land Capacity Task Force. 1995. *Findings and recommendations of the King County Land Capacity Task Force.* Seattle. Submitted to Growth Management Planning Council of King County. November.

King County Office of Budget and Strategic Planning. 1997. *Draft commercial and industrial land capacity report.* Seattle. October.

Portland Metro, Data Resource Center. 1990. *Metro employment density study.* Portland.

Puget Sound Regional Council. 1995. *VISION 2020 (1995 Update).* Seattle: Puget Sound Regional Council. May.

Puget Sound Regional Council and University of Washington Center for Community Development and Real Estate. 1998. *Industrial land supply and demand in the central Puget Sound Region.* Seattle: Puget Sound Regional Council. February.

Real Estate Economics. 1984. *Industrial land market analysis, Kitsap County, Washington.* Bellevue, Wash. Prepared for the Economic Development Council of Kitsap County. July.

Rhine, Richard, MPA. 1996. *Summary reports for data submitted in section B of CTED Growth Management Services Annual Reports.* Olympia, Wash. Report for Washington State Department of Community, Trade, and Economic Development, Growth Management Services Division. November.

Snohomish County Department of Planning and Community Development. 1985. *Business and industrial land survey.* Everett, Wash. December.

Snohomish County Department of Planning and Community Development. 1994. *Employment land capacity analysis for unincorporated Snohomish County.* Everett, Wash. April (revised June 1995).

Therrien & Price, L.L.C. 1995. *Industrial land inventory of Snohomish County.* Seattle. Report for Economic Development Council of Snohomish County. June.

COMMENTARY: METHODOLOGICAL LESSONS FROM PUGET SOUND LAND MONITORING

Scott A. Bollens

The PSRC/CCDRE study, *Industrial Land Supply and Demand in the Central Puget Sound Region,* is similar to studies such as those of Portland and Montgomery County in that it deals with data from multiple jurisdictions. However, this study is notable for its focus on industrial rather than residential land supply. The method relies on "concentrations" of contiguous planned industrial land (developed or undeveloped) at least 25 acres in size as the primary units of analysis for investigating the adequacy of industrial land supply in the Seattle region. The database was assembled by using parcel data at times (such as the identification of vacant and redevelopable land and the determination of parcel size), as well as subarea- or jurisdiction-level data for such attributes as infrastructure proximity and availability, and environmental factors. Within each concentration net industrial land supply was estimated based on vacant and redevelopable parcels, minus (1) land unavailable owing to environmental constraints (wetlands, flood areas, and steep slopes) (2) a discount for future street rights-of-way, and (3) a discount to account for "uncertainties."

The authors identify clearly the study's limitations. First, it is a one-time snapshot rather than a monitoring system that can evaluate changes in the supply-demand balancing act over time. Second, although data are compiled at the concentration level, supply-demand analysis is done mainly at the regional level, and not for specific concentrations. Third, it analyzes aggregate regional supply-demand for industry generally, and not for specific sectors. Fourth, according to the authors, their not using parcel data created certain problems, such as the inability to screen for parcel shape (important to industrial developers), to fine-tune environmental constraint assessments, or to perform spatial analysis of the potential for reassembling parcels.

Given these constraints, the case study highlights issues and characteristics of supply inventory systems. First, it brings to the fore a critical issue in monitoring systems: the variety of ways of defining land supply and the related terms used. Criteria to define land supply include land vacancy, lack of environmental constraints, proper zoning, availability of infrastructure, presence of land currently on the market, and economic feasibility of development. In various systems around the country, terms such as "available," "developable," "buildable," "feasible," "marketable," and "suitable" are used to mean different things at different times (one system's "available" supply is another's "buildable" supply). Absent some nationwide uniformity of terms used, it becomes important to state and clearly articulate what a particular label means and to explain the assumptions being used in determining an overall adequacy of land supply.

In the Puget Sound case, "net land supply" is based on land vacancy and lack of environmental constraints (with an allowance for space needed for future roads). Later, analysts evaluate this net supply in terms of its land "readiness." Readiness is determined on the basis of availability of infrastructure, presence of contaminants, and parcel sizes. Interestingly, the report concludes that supply exceeds projected demand by a factor of three. However, this estimate is based on "net land supply," not the more constrained estimate of "ready" land. That the two estimates will be significantly different is shown by the fact that 33 percent of net land supply is not currently serviced with adequate infrastructure. In addition, information provided by brokers in mid-1997 indicated that only approximately 20 percent of net supply was then on the market. These factors point to the variability among estimates of available land, depending on how one views what land is truly available. Without attention to infrastructure availability and private sector considerations, Puget Sound planners may be overestimating available industrial land. On the other hand, the fact that statewide historical data has shown that about 40 percent of industrial development occurs on land planned for other than industrial uses indicates that planners may be underestimating current supply.

Furthermore, the criterion of infrastructure availability is used differently in Puget Sound and Montgomery County. In the first case, it is not part of land supply calculations; in the second case, it is an integral part of such reckoning. This contrast is due to the different time horizons involved. The

Puget Sound Regional Council is looking at the 20-year supply-demand and assumes (likely with good reason) that infrastructure will become available as needed. Montgomery County, on the other hand, examines supply-demand on a yearly basis, and the availability of infrastructure is a guiding determinant in the county's growth management program, with project approvals dependent on it. This illustrates how the definition of land supply is dependent on planning goals and indicates that monitoring system parameters should be built around intended outcomes of growth governance.

Second, how land supply is defined relates to the differences in views held by the public and private sectors. Governments usually view land supply in such categories as vacant, physically unencumbered, serviced by infrastructure, and properly zoned. For example, state law in Oregon (Oregon Administrative Rules 660-07-005) specifies that a "buildable land inventory," required of all cities, must account for land that is "residentially designated vacant land and . . . redevelopable land within the Metro Urban Growth Boundary that is not severely constrained by natural hazards or subject to natural resource protection measures." The Department of Community, Trade, and Economic Development in the State of Washington recommends that the potential availability and capacity of infrastructure be included as important criteria and should, in addition, be documented in deciding on adequate land supply (Enger 1992).

Private interests, on the other hand, view land as "truly" being in the supply and "developable" only if it can be purchased and is economically feasible to develop. Land supply estimates based solely on vacant and underutilized classifications, zoning, services, and physical constraints tend to overestimate the supply of "truly buildable" land, inasmuch as some of this land will not be on the market or not be economically feasible to develop. A report by the Real Estate Research Corporation (1982) asserted that land otherwise appropriate for development may still be held from development for reasons of investment or personal use. On the other hand, land supply estimates that include market availability and feasibility tend to underestimate land supply as a result of incomplete information, uncertainty, and the dynamics of these market dimensions. Generally, public-sector measures of land supply are adequate for the long-term and private-sector measures address short-term conditions.

Although it appears to be beyond the ability of local government to assess the market dimensions and feasibility of development, jurisdictions should strive to refine the usual public-sector definition of land supply so that more accurate, short-term estimates of land supply are possible. A Stockton, California, study used a stratified random sample of landowners to provide a basis for estimating market availability, and compared housing prices in various locations with the costs of producing housing to determine economic feasibility (Gruen, Gruen and Associates 1982). The Puget Sound Regional Council could, for instance, use the broker-supplied information generated by the study to qualify its estimates of land supply. Linking the monitoring system to real estate Multiple Listing Service (MLS) files could also provide an estimate of how much of the land supply is on the market at a given time, as compared with projected demand for the same period. Similarly, the identification of "ripe" parcels—those owned by development firms, builders, or speculators—can add to the validity of buildable land supply estimates.

Land price monitoring could also be incorporated into the land supply database as one of the analytical means of determining whether a given land supply is adequate relative to demand. If a system is parcel-based, data on sales date and price could be routinely entered by the assessor's department. Scientifically sound methodologies for measuring annual changes in land prices could be developed. Land price indices should focus on standardized parcel characteristics so as to control for the effects of location, infrastructure, and neighborhood externalities. Price indices can be developed either from the sales data component of the monitoring system or by periodically surveying real estate experts in the area. Sales and assessment data monitored regularly can be used to construct jurisdiction-wide land value maps that could pinpoint those subareas undergoing accelerated price inflation.

Overall, the Puget Sound case shows both the utility and limitations of using units of analysis larger than parcels. The existence of significant data at jurisdictional and subarea levels allowed planners to compile this information with moderate cost and complexity. On the other hand, without primary parcel data, the analysis necessarily stayed at the level of subareas, without the ability to analyze parcel shape, to identify the potential for land consolidation, or to analyze environmental constraints in more detail.

The Puget Sound Regional Council is now moving to incorporate parcel data into its land supply accounting. This will require further multijurisdictional cooperation in creating a reliable and internally consistent database. Time, cost, and complexity will surely increase. Nevertheless, it appears that a parcel-based system may be appropriate with industrial land. The problem of changing parcel geography (identified earlier in Chapter 3) appears to be less a problem with industrial than residential land owing to more stable boundaries, although problems with subparcel heterogeneity remain. In the Puget Sound case, the size of the database (about 4,000 parcels in a land supply of roughly 21,000 acres) may moderate concerns about time and cost in building a parcel-based GIS for monitoring that supply.

References

Enger, Susan C. 1992. *Issues in designating urban growth areas.* (Parts I and II). Olympia, Wash.: State of Washington, Department of Community Development, Growth Management Division.

Gruen, Gruen and Associates. 1982. *The need for housing and additional land for residential development in Stockton.* San Francisco.

Real Estate Research Corporation. 1982. *Infill development strategies.* Chicago: American Planning Association.

COMMENTARY: ASSESSING DEMAND FOR INDUSTRIAL LAND

William Beyers

The case study appears to be an excellent piece of research, which carefully documents the land zoned for industrial uses and the current acreage in industrial use, and estimates future demand for industrial land to the year 2020. Using the PSRC forecasts as a basis for these demand estimates, the authors conclude that there is considerably more land currently available for industrial uses than will be needed in the year 2020. However, they also note that the geographic locations of much of this land are relatively remote from current major concentrations of industrial land use.

This commentary relates to both the study and the need for additional research on aspects of the demand for industrial and, more generally, employment land in our growing regional economy.

First, the authors have uncritically accepted the PSRC employment forecasts. Yet uncertainty surrounds these estimates, as illustrated by the current decision of Intel to shut down a portion of its brand-new facility in Dupont, Washington, because of sluggish demand for chips. Similarly, the recent history of employment fluctuation at the Boeing Company will undoubtedly be repeated in the future, although it is impossible to predict the exact cycle of fluctuation. Alternative forecasts of industrial activity could be pursued, so as to get high or low estimates, in the spirit of many economic forecasts.

Second, the level of churning and redevelopment is not captured in a study of this type. The frequent scenario of small firms outgrowing the space they occupy, thus requiring a move to gain new space, produces vacancies to be filled by other establishments. The "coverage ratio" may rise for particular parcels beyond that assumed in the study—another parameter that could be evaluated for optional acreage demands.

Third, the relatively peripheral location of much available industrially zoned land may, in fact, be a blessing, not an adverse situation as implied in the study. Within the context of *VISION 2020*'s system of major employment centers located close to residential developments (Puget Sound Regional Council 1995), it may be that the region should support development of these now relatively exurban or suburban centers to reduce long-distance commuting into the current industrial core by additional workers in the industrial sectors.

Finally, although this is a good study of the demand for land by one sector in the regional economy, we need similar research focused on other growing sectors—in what I have referred to as the "New Economy." The vast bulk of future employment growth will not be in the industrial categories, but in various service industries that constitute the New Economy—documented in PSRC employment forecasts. The New Economy is a complex network of producer services, entertainment and multimedia industries, and proprietors working sometimes from their homes. It includes professional and technical occupations, as well, tied strongly to the higher education system not only for basic training, but for ongoing reeducation and support. Strongly reliant on telecommunications, it also depends on person-to-person interaction, which places demands on the regional transportation system and the air travel network. It relies on small-package courier services and can operate successfully in either high-density office environments or dispersed locations.

What do we know of the demand for land in the region as related to the New Economy? Has *VISION 2020* adequately anticipated the demands on land, infrastructure, office and work space, and settlement patterns for the New Economy? PSRC has to assess whether the land supply and demand issues related to the growth of the New Economy are adequately documented, and then decide whether to undertake the appropriate analyses to complement this study of industrial land supply and demand.

Reference

Puget Sound Regional Council. 1995. *VISION 2020, 1995 Update.* Seattle, May.

Part III

Thematic Issues

6

Method and Technical Practice in Land Supply and Capacity Monitoring

Ric Vrana

The current generation of parcel-based geographic information system (GIS) technology is but the latest in a long line of measurement and representation methods for the inventory and management of land resources. In a curious way, settlement patterns and the geographic description of land ownership are intricately bound with the economic and cultural determinants of historical and technologic eras. Of course, this is not to say that prevailing mapping technology determines land use any more than land ownership patterns create mapping technology, but, to step back just a little—can we not discern a kind of dialectic momentum to these geographically grounded human endeavors?

Metes and bounds surveys, French colonial long lots, and the Jeffersonian rectangular survey system are all ownership and, hence, land use patterns reflected in the North American landscape (Jordan 1982). In 1930s Britain, the availability of large-scale topographic base maps enabled local volunteers to carry out the nationwide Land Utilization Survey (Stamp 1940; Board 1968). At about that time, the "unit area" approach to land use characterization of the Tennessee Valley Authority project resulted from the introduction of air photography to land use analysis in the United States (Hudson 1936). Some

Monitoring Land Supply with Geographic Information Systems, edited by Anne Vernez Moudon and Michael Hubner ISBN 0 471371673 © 2000 John Wiley & Sons, Inc.

149

years later, the records automation environment of urban renewal in the 1960s spawned the Standard Industrial Classification (SIC)-linked *Standard Land Use Coding Manual* (URA-BPR 1965; Clawson and Stewart 1965). The need for making use of high-altitude aerial photography, and later satellite remote sensing, resulted in the land use–land cover classification system employed by Anderson et al. (1976) for use in the Geographic Information Retrieval and Analysis System (GIRAS) resource inventory and land cover mapping projects (Mitchel et al. 1977).

Thus, in viewing the current state of the art in monitoring land use, supply, and capacity, we recognize that the available technical means for analyzing land information influences the basic methodology for managing it. Changing societal mandates, such as those directed at improved urban growth management, tend to push the prevalent methodology until more technical innovation is required. In this context we can review technical advances that have made or will soon make parcel-based land information systems ubiquitous tools for examining land use and land supply. Having provided access to so much land information on such a detailed scale, these systems raise institutional expectations for managing ever more detailed and timely data. Of course, we do not slight the important influences of successive rounds of organizational responses to technological development addressed elsewhere in this book (see Chapter 7). They also contribute to the revised mandates for monitoring land supply and plan compliance. Together these forces bring new technical barriers into focus, eventually to be reconciled in the next round of methodological advances.

Within this broader dialectic we can look at a short cycle of technological evolution, dealing with the introduction of GIS into land use planning, roughly in the mid-1980s, up to the present. This chapter first acknowledges some key technical advances that brought us to the current configuration of systems. Next, the requirements of growth management applications suggest some important technical challenges to be overcome. A number of suggestions for revised land monitoring methods emerge as feasible practices, assuming we are meeting the technical challenges they may pose.

ADVANCES TO THE CURRENT STAGE (OR HOW WE GOT TO WHERE WE ARE)

Prior to the adoption of geographic information systems, tabulations of land use from parcel-level inventories may have been compiled onto a regional base map, but, more often than not, these inventories were completed only for subarea neighborhood studies. Comprehensive land use maps for entire municipalities were thus likely to be out of date or to have widely varying dates of currency. Urban development at the fringes of cities was reflected in the official record on assessor plat maps, which were updated by hand. Coordinating objectives and zoning regulations with development applications, utility provision, building inspection, and land supply analysis eventually required an automated land information system and various technical and organizational efforts to modernize land records management. By the early 1980s, it was apparent that land data had to be integrated from multiple sources and organizations and that geographic information system technology would be useful to facilitate this, as embodied in the concept of the "multipurpose cadastre" (NRC 1980, 1983; McLaughlin 1984).

In terms of land records and geographic information technology, we can identify three overarching technical developments that spurred the development of this concept: (1) topological overlay, (2) relational data models and associated database functionality, and (3) advances in user interfaces and software portability.

First, the development of vector data structures for topologically representing ownership parcels as polygons was the technical means by which the integration of the many layers of the multipurpose cadastre concept could be implemented. Although the earliest land information system proposals did not include geographic layers independent of the property parcel, it soon became apparent that there was a wealth of mapped environmental and socioeconomic data that could be related to property parcels by topological overlay. Dissenters, such as Boyle (1987), argued that most of the needs for data integration could be met by representing parcels as centroids intersected with *point-in-polygon* methods to link to other spatial data. There may be specific applications for which this is a parsimonious analytical solution, although it relies on the rasterization and registration of lot lines as the visual confirmation that these are, indeed, parcels. The precise locations of boundaries are known, however, and surveyors and engineering personnel charged with maintaining the maps usually have reason enough to seek the efficiency of automation. Ultimately, it may be the need for timely and cost-efficient cadastral mapping updates that have driven the creation of digital parcel maps for many jurisdictions.

Second, the development of relational data structures integrated into GIS not only facilitated the overlay capability of topological data; it provided the means to link tax assessment and other databases with the parcel base map. This gave assessors, along with planners, an institutional stake in the development of jurisdiction-wide GIS. With the widespread success of this database management approach came standards for query languages, export formats, and, eventually, easy integration of the GIS data into other common office software applications.

Third, improved user interfaces for GIS software, the migration of systems of minis, then to PCs, and more recently the interconnectivity of PCs on local and wide-area networks, further expanded the usefulness of the parcel-based GIS for a wider range of participants involved with planning, managing, and developing land resources.

These technical advances were evolutionary, taking place in many iterations, among quite a few software vendors, not all of which survived to support the next round of innovation. But the net effect is that in parallel, these developments have expanded the user base for parcel information and made it easier to collect, maintain, and visualize spatial relationships involving land uses and developable areas. In turn, an increasing number of planners, assessment officials, developers, and engineers required applications for parcel-based land data. Not only did this boost the efficiency of existing mapping and analysis, but the increased volume and resolution of land data suggested new ways of monitoring the development of land supply in an increasingly "regulated landscape" (Knaap and Nelson 1992).

CHANGING MANDATES FOR LAND USE MONITORING UNDER GROWTH MANAGEMENT

Communities employ planners to develop plans and alternatives to plans, and to monitor the evolving implementation of a plan. Most prominently, this includes tracking the finite store of land available for development. This concept is embedded, for instance, in the Godschalk et al. (1986) formulation of "land supply monitoring systems." Similarly, Chapin and Kaiser state that "systematically tracking land use change is the single most critical activity in maintaining a credible land use information system" (1985, 203). Tracking changes requires some initial inventory, an agreed-upon means by which change is to be measured, and some subsequent data collection for comparison at a later time. Increasingly, this takes place with respect to what has come to be called growth management.

Growth management has become the legal and regulatory framework for planning the disposition and supply of land in a number of states and metropolitan regions across the United States. De Grove (1992), Easley (1992), and Stein (1993) each provide comparative analyses of state systems. These and other authors point out that common to most growth management strategies are the three interrelated goals of consistency, concurrency, and compact urban form. Land data that support the goals of consistency, concurrency, and compact urban form must be available to

1. ensure comparability with planning in other departments of the municipality and with neighboring locales;
2. be linked to updated records on the extension of urban services and facilities; and
3. monitor progress toward target densities and quantify developable land supplies.

In addition to these requirements for land data, growth management has given rise to the expectation that planning should become open to public participation. In many localities, land use planning indeed has come under much public scrutiny. Technological innovation can improve public access to spatial data and information systems for analysis. However, the technology is only as good as the data and the institutional arrangements for its acquisition and maintenance over time.

A certain level of technological advancement thus contributed to the feasibility of growth management planning. The adoption of growth management objectives, in turn, constitutes a significant change in the mandate for land use information. Monitoring change requires a shift from a static view of land use data collection to the representation of land use transition. Couclelis (1991) considers this mandate for understanding planning space as "relational," to be somewhat at odds with the reductionist inventory treatment of space as a "container." That is to say, in the context of capacity analysis, mixed-use development objectives, and growth management, land use must be viewed as contingent on a variety of temporal processes and not simply as an attribute describing a parcel. Thus, it is no longer sufficient to periodically inventory land use characteristics. To monitor the implementation of planning objectives, we must become knowledgeable of how parcels are used in relation to each other. To know how the store of available land is holding up, we must be able to track land transactions that affect use and how the resulting changes affect existing and potential land use elsewhere.

METHODOLOGICAL RESPONSES TO THE NEED FOR LAND SUPPLY AND CAPACITY ANALYSIS

Conventional approaches to land supply and capacity analysis accommodate urban growth by a combination of identifying vacant, or available land and, where the data are sufficient, monitoring the conversion from this state to residential, mainly single-family use. Bollens and Godschalk (1987) prescribe a somewhat sophisticated, multiple-step analysis to exclude lands restricted by environmental factors from being considered "available" and further recommend that some approach to identifying economically viable lands for residential conversion be identified. Adjusted available supply and capacity are construed not only to be dynamic over time, but also to vary with respect to market factors such as housing demand, land price, and design decisions to allocate growth in special areas—such as Seattle's "Urban Villages" or Portland's "Functional Design Areas," which attempt to implement various schemes for mixed residential and commercial uses. Methods for estimating land supply and capacity therefore develop out of the need for planners to refine this analysis, given better assessment and land use data.

Portland Metro represents an interesting case of a locality grappling with these and other methodological responses to growth management. Metro is the regional metropolitan planning organization for a three-county area comprising most of metropolitan Portland. Under the basic framework of Oregon State land use law, the aim of comprehensive planning for the urban area is to establish an urban growth boundary beyond which certain land use transformations are not to occur. "Greenfield" developments at or near the urban fringe are typified by the transition from extensive agricultural or resource uses to certain residential, strip commercial, or industrial park uses. Portland Metro operates the Regional Land Information System (RLIS), a parcel-based land information system containing more than 400,000 tax-lot records linked to assessors' records of three counties, a cartographically correct street centerline file that registers with the tax-lots, and various environmental and natural feature overlays. A long-range planning process has established a framework for managing land resources in the face of the estimated doubling of the region's population over the next 50 years.

In addition to conventional land supply factors, a number of transportation, open space, and environmental issues constitute the "Performance Measures" by which the planning agency reports progress toward meeting growth management goals. Officially, "performance" consists of the following nine categories for which measures are to be devised and progress reported to the Metro Council (Portland Metro 1997) (author's comments in parentheses):

1. Vacant Land Conversion
2. Housing Development, Density, Rate, and Price (a big issue)
3. Job Creation (not clear about specific sectors)
4. Infill and Redevelopment
5. Environmentally Sensitive Lands
6. Price of Land
7. Residential Vacancy Rates (but not commercial vacancy rates)
8. Access to Open Space
9. Transportation Measures (mostly in terms of reduction of vehicle miles traveled)

The parcel-based information system must be used to monitor the region's compliance with specific design targets in several areas at once. Internally, a number of specific indicators will likely be tracked over time to derive the reported Performance Measures. These include target benchmarks for

- the supply of land available for development within the urban growth boundary (the state mandates a 20-year supply be maintained);
- housing and employment densities summarized by Functional Design Areas (FDAs, also called Plan Designation Areas) and subareas, and for the urban area as a whole; and
- the mix of land use types that will characterize the FDAs. These include areas designated as "central city," "regional mixed-use centers," "inner" and "outer urban residential neighborhoods," "town centers," "light rail

transit station centers," and similar designations characterized by both density and land use mix.

The target densities are greater than the current average for existing areas that will build out according to criteria for FDAs. Politically, it is proving extremely difficult to expand the region's urban growth boundary, and the expectation is to achieve the mandated targets mainly through infill and redevelopment—thus the emphasis on mixed-use centers. Infill is defined as new construction on vacant land within areas that are already largely developed, even when a subdivision or lot line adjustment is required to permit it. Redevelopment, on the other hand, assumes the replacement of one use by another, such as when a structure is torn down and another one erected. Both of these objectives are to be driven by the economics of land conversion, aided by appropriate zoning and building ordinance revision. Although the restriction of urban expansion is expected to be the engine of such conversion, there is a real—some say inevitable—danger that land prices will soar to unaffordable levels (Mildner et al. 1996). Because this trend would be counterproductive to affordable housing goals, and because an overheated land market would weaken the political will to carry out the long-range plan, monitoring this process of increasing densification becomes an even greater priority for Portland Metro.

Maintaining a Reserve Supply of Vacant Lands

Oregon State law requires Portland to have a 20-year reserve supply of developable land inside existing urban growth boundaries. The baseline having been established in the late 1970s after the implementation of statewide land use planning, it is now time to reassess this boundary in terms of the supply objective. The issue is being approached in two ways. First, it is necessary to gather a fairly precise inventory of existing vacant lands and, beyond that, to determine which of those vacant areas are likely to respond to the target growth densities. Second, some of the more permeable portions of the urban growth boundary are being scrutinized and selected adjacent areas are being identified as "urban reserves" into which the boundary is likely to expand.

The vacant lands inventory is revised annually, consisting of current air photo reconnaissance compared with the tax-lot map to identify vacant portions of individual parcels large enough to support subdivision and infill. The results are then compared with a compilation of building and development permits to track how those lands have undergone conversion in the past year.

Development permits are periodically compared against the tax lots to identify change. In a 1995 internal study, Metro found that up to 29 percent of the previous year's new residential permits had been assigned to parcels that were not assessed to be vacant. Three scenarios could account for this apparent discrepancy, each requiring additional data development or analytical methods to track (Vrana and Dueker 1996):

1. Some parcels that were available for development were falsely characterized as occupied in the assessor's land use file. *Solution*: Check the parcels against aerial photography to confirm coding error. Update land use records if necessary.
2. Development occurred on a vacated street or other right-of-way or easement. *Solution*: Access the development permit records and see whether such actions were taken by the local jurisdiction.
3. Development occurred as a result of an adjustment to the parcel. This is the most likely possibility, fitting neatly into the definition of infill by subdivision. *Solution*: Locate records of subdivision and lot adjustments and check for activity.

The detailed level of monitoring implied in the preceding paragraphs is necessary to distinguish new development from infill. Knowing when a land use action constitutes infill is important, because it can help to determine whether the residential or employment density is moving in compliance with plan targets, and it can establish the rate at which the available land supply is decreasing.

Redevelopment

In addition to infill, a certain amount of transition can be expected to consist of redevelopment from existing uses. Redevelopment is attractive in this situation because it can lead to target densities without actually decreasing the available supply of developable vacant land. At the same time, however, redevelopment does not guarantee greater densities. Being able to distinguish redevelopment of existing uses from infill is important in order to monitor redevelopment to see whether target densities and land use mixes are being achieved. By definition, redevelopment can be considered any transition between land uses, one of which is not "vacant." In practice, it is not difficult to determine when redevelopment is occurring. The trick, however, is to formulate an estimate of where it is likely to occur and how much urban growth will be absorbed by redevelopment of existing uses. The conventional approach is to compare a site's (or parcel's) building value to its land value. The lower the building value is relative to the land, the more likely a property owner is to invest in new construction to increase the economic productivity of the parcel. Problems inherent in this approach include (1) establishing the cutoff value that is likely to be considered economically feasible to consider the parcel as part of the "adjusted" capacity, (2) establishing market values, given out-of-date and sometimes understated assessed values, and (3) sorting out the interdependence between land value, building value, and location-specific attributes.

With respect to the first problem, Portland Metro derived an estimate of land likely to be redeveloped by using grid neighborhood functions with a GIS. After eliminating parks, designated open spaces, and vacant and environmentally restricted land, the tax-lot map was converted into 52-by-52-foot grid cells, each inheriting the building and the land value of the parcel occupying most of the cell. Each grid's building-to-land-value ratio could then be compared with a moving average of surrounding cells. The grids were realigned with parcels, and, depending on the cutoff value percentage difference, estimates of total redevelopable land were determined as a function of the surrounding neighborhood. As clever a technical solution as this is, however, it derives only an estimate of likely redevelopment. It does not take rising (or sinking) market property values into account over time. It ignores the fact that some land uses are more likely to be redeveloped than others and that some parcels have a transportation or other locational advantage over others. A model incorporation these variables would not be impossible to implement, but it is not known how significant its advantages would be; thus, this is an area that should be investigated.

Housing and Employment Densities

Housing and employment densities are moving targets. Although the long-range planning goals define specific densities for subareas and FDAs, progress toward meeting these goals must be measured and reassessed periodically. Over time, in fact, the target goals may be adjusted, but this can happen only on the basis of some estimate of the degree to which they are incrementally being met. Therefore, it is necessary to develop measurements of compliance in the form of static densities for the baseline and for any subsequent temporal sampling. Change measures comparing the baseline, or any other archived rate, against a subsequent time period are also needed. This difference can be measured either by counting land use or parcel-altering transactions during a period or by measuring the densities at two points in time and calculating the difference.

Another issue in constructing "status" and "change" variables is the selection of the areal unit for which quantities will be measured. Regionwide, a measure of current development practices is often attempted by using new subdivisions as the unit of measure, but comparability over long time frames makes this problematic. Furthermore, this approach reinforces a somewhat myopic preoccupation with new single-family construction at the expense of all other categories of land use. Block-level comparisons are unduly small aggregates and quite variable in size. Traffic analysis zones (TAZs) also suffer from some degree of size inequality, but because they are inputs into transportation modeling analyses, they should be considered, especially for employment densities. Census block groups and/or tracts provide easy comparison with socioeconomic variables. Where appropriate, the neighborhood may be a reasonable unit of measure, although this is more ambiguously defined in some communities than in others. Over a metropolitan region, where zoning and plan implementation are mainly the responsibility of local jurisdictions, it is advisable to extract compliance measures for those jurisdictions. Finally, in the case of Portland, where FDAs are employed as a strategy for encouraging certain mixed-use combinations at specific densities, these too, should be summarized by housing and employment density.

Summary: Monitoring Status and Change Variables

The methodological response to the combination of better data resolution, increasing analytical capabilities, and a changing planning mandate requires ongoing monitoring. In Portland this monitoring consists of periodic measures of housing/employment balance, land capacity, and use mix for designated FDAs. Moreover, these measurements are not just static estimations for fixed units of time. They should include an established baseline inventory and then the calculation of rates of change, or rates of compliance to plan targets, over the life of the long-range plan. Periodically, based on these measurements, we can expect both the targets and the methodology to change, again as the result of another round of technological innovation and changing demands for information.

Table 6.1 summarizes a number of proposed indicators to track the change variables associated with the static measurements. This list does not exhaust

TABLE 6.1   Status and Change Variables to Monitor Plan Compliance

Status Variable	Change (X) Variable	Unit of Measure	Comments
HOUSING DENSITY			
Gross density (GD) = # housing units (HU)/area	$XGD(t_a - t_{a-1})$	FDA	FDA = Functional design area
Net density (ND) = # HU/res. area	$XND(t_a - t_{a-1})$	Jurisdictions	
Actual/allowed by zone (A/AZ)	XA/AZ	Zone, FDA	
Actual/Allowed by FDA	XA/XF	Jurisdictions	
SINGLE-FAMILY DENSITY			
Gross lot size (GLS) = area/# lots	$XGLS$	FDA	
Net Lot Size (NLS)	$XNLS$	Subdivisions	Net excludes right-of-way
HOUSING MIX			
% Single family (SF)	X % SF	FDA	
% Duplex	X % Duplex	Jurisdictions	
% Multifamily (MF)	X % MF		
LAND USE MIX			
Jobs/Housing (J/HJ) = # jobs/ (#HU + # jobs)	XJ/HJ	FDA, TAZ	For transportation modeling
% Public = Pub LU/Area − (r.o.w. + unbuildable) commercial/residential (C/R)	X % Public XC/R	Jurisdictions, FDA	

Source: Adapted from Vrana and Dueker (1996) and Vrana (1997).

the possibilities, of course, but is intended to point out the range of land use measurements to be performed in such a planning model.

THE NEXT ROUND OF TECHNICAL IMPEDIMENTS

Technological advances in computing architecture, processing speed, software user interfaces, and networked data sharing are moving GIS applications along evolutionary paths parallel to those of the broader community of computing applications in general. These considerations have had an impact on the way planning is, or will be, conducted. For example, the development of standardized embedding and linking functions that include GIS functional modules encapsulated in regular business applications (OLE, ESRI Map Objects, etc.) already provide new customized tools for analysis. These innovations and, not incidentally, the development of web-delivered GIS functionality via Internet Map Server and similar technology, are expanding the availability of GIS to the general public.

In addition to these "normal" technical issues in applications development, three broad areas of technical impediments relate specifically to the land analysis question. They can be classified under the general heading of "data model issues," although the first is more generally a process model of institutional records keeping. These impediments generate bottlenecks in analysis and define areas of inquiry for both GIS and urban planning researchers:

- Incorporation of historical or legacy data on land use states, and development of better methods for capturing and storing the land use transactional record
- Improved definitions of land use categories, explicit modeling of use attribute relationships, and the means to translate one characterization scheme into another
- Models for the representation, query, and analysis of spatiotemporal data in GIS

Overcoming these impediments will facilitate the methodological responses considered in this chapter thus far. This is not to say that the methods described cannot already be implemented. It just takes a while to learn how to do it cleanly and efficiently.

Legacy Data and the Transactional Record

Establishing a record of land use transition and consumption requires recourse to historical data. For quite a number of years, planners have dreamed of tapping into the continuous transactional record of subdivision plans, permit applications, building inspections, or business license applications for a continuous updating of land use, supply, and capacity (Craig 1980).

Converting old land records, where available, into currently usable ones can be a daunting task. It probably should be done by an enthusiastic student with a detective's instinct, or a seasoned Zen master. In addition to the normal problems related to uncovering old data, historical cadastral data may be archived in formats not normally encountered with current GIS tools.

At its heart, however, the problem is less one of data conversion than one of streamlining the institutional data work flow. This has an organizational, data sharing component as well as a technical one. Transactional data systems for in-house permitting and review are currently under commercial development and are beginning to appear in larger planning agencies. Engineering these systems to work in conjunction with parcel-based GIS will require decisions about when, in the process of a development application, for instance, a land use event should be represented in the GIS database and to what spatial or temporal entities that event belongs. That is, the task of rationalizing models of the transactional data work flow in an organization will have to be folded into the task of defining classification and spatiotemporal models as described in the following sections.

Meanwhile, provisional evidence shows that it will take major effort to capture a complete transactional record. This author's experience with a land use transition study for the city of Kent, Washington, illustrates a common situation (Vrana 1997). In this case, the transactional record and the reconstruction of a current land use survey using a baseline survey from ten years earlier did not reveal most of the change evident on-site. In particular, the records failed to reveal changes in business operations, remodels and minor use changes, and changes in the configuration of multiple uses. The transactional record was searched to document observed change in the resurvey. In one of the surveyed sections (1 square mile) consisting overwhelmingly of residential land, approximately 76 percent of all cases involving land use change could be documented. In a neighboring section, composed mostly of commercial and industrial uses, only 42 percent of the cases could be documented. The difference reflects the fact that changes in residential land use often involve lot line adjustments and, possibly, subdivision, whereas most retail and other commercial turnover involves a change in tenancy but not necessarily a change in the parcelization record. It seems that the precise delineation and ownership of land parcels remains a looming preoccupation in modern urban culture. Moreover, although greenfields conversion of vacant land is generally well documented, less attention is paid to keeping business licensing and office park tenancy current.

Taxonomies for Land Use Classification

"Land use" differs from "land cover" in that it must capture some meaning of utilization not independently observable from remote sensing platforms. In contrast to land cover classifications (see Anderson et al. [1976] in respect to classification levels I and II), land use classifications are derived from a complex set of social, economic, and cultural relationships and cannot be deduced from a spectral signature. Image-based sensing is useful and necessary in the delineation and updating of urban use classes. In addition, satellite remote sensing data are often well tailored for modeling and simulation involving land use transition in wider areas, as demonstrated recently in Logsdon et al. (1996). The main issue to remember with respect to image analysis and land use, however, is that there is not only a many-to-one relationship between land use and land cover, there is also a many-to-one relationship between land cover and land use. For example, a single land use class, "Multifamily Residential," may be characterized by several land cover types that differ in vegetation canopy or reflective roof and pavement surfaces. At the same time, of course, a given roof reflection pattern can be evidence of commercial-retail, light manufacturing, or multifamily housing. In practical terms, this means that land cover analysis with remote sensing can be used to confirm the existence of change (e.g., a decrease in vacant supply) or rule out the possibility of a specific land use change hypothesis; other means are necessary to confirm the existence of land use instances "on the ground."

In most cases, "ground truth" for land use categorization derives from a functional consideration of a parcel's role in the economic or aesthetic life of the city. Coding schemes for classifying urban land use are largely a reflection of existing (or neglected) economic inputs. Because economic activity is categorized in terms of industrial sectors, land use data must be compatible with, or at least directly linked to, data on economic activities. In addition, nonindustrial sector uses, such as residential, recreational open space, and various cultural uses, must be specified within the same classification framework.

Furthermore, a land use code that serves this purpose should be easily queried and aggregated for generalization. It should be extensible, defined in ways that allow new uses to be coded as logically "near to" similar uses. Taken together, these requirements have produced hierarchically modeled taxonomic schemes for land use characterization, such as in the *Standard Land Use Coding Manual* described and explained in Clawson and Stewart (1965). However, the hierarchical structure can mask functional similarities between parcels coded in quite different categories. For example, gas stations with convenience food outlets are in a different class from convenience food stores with gas pumps. At some point in the resolution of use, field delineation by decision rules becomes an arbitrary exercise. Although delineating food outlets with or without gas service is not trivial, it is merely an artifact of the classification scheme. Standard Industrial Coding (SIC) links may be invaluable for referencing commercial ownership and land use class, but widely different SICs can be recognized as essentially identical functional uses on the ground.

In addition, multiple use presents at least three conceptually separate issues. First, multiple uses may vary in time. For example, diurnal or seasonal changes in the uses of parcels are not at all uncommon. Second, two or more firms, collectively engaged in dissimilar SIC activities, can be tenants on the same parcel—a prevalent situation in industrial or corporate parks. Third, many parcel records reflect multistory buildings where the land area is simultaneously engaged in multiple uses, even if building floor space per level is not. It is important to note that this characterization is expected to increase as a proportion of all use parcels with the implementation of growth-management-induced mixed-use development strategies. Perhaps certain combinations of land uses may themselves constitute land use classes, thereby introducing added complexity in the many-to-one relationship between land use and SIC classifications. Finally, "multiple use" is not identical with "mixed use," as has been described in some planning objectives: Whereas multiple use refers to the land use condition of a parcel or tax lot, mixed use concerns the set of land uses existing in close proximity in a given area, where individual parcels can be in multiple or single use.

The most easily obtained data on land use for parcel-based urban land information systems are assessment and taxation records that can be linked to tax-lot maps. Assessors normally include a land use classification as part of the data collected on individual parcels for the purposes of assigning value, tax rates, or determining exemption status. However, "land classification for assessment" does not necessarily correspond with activity classes recognizable on the ground or, indeed, with the many variations of economic role in the urban system. For example, it is not uncommon for assessment classifications to make detailed distinctions between single-family building styles while at the same time lumping all commercially zoned land into a single category. Therefore, when assessor's data greatly generalize actual land use, they sometimes do so to the point of establishing misleading information when applied in a planning or development context.

Moreover, both hierarchical economic sector and tax assessment classification schemes may be applied differently among jurisdictions. This can be a particular problem for adjoining jurisdictions collectively engaged in planning or land supply monitoring. Portland Metro again illustrates this point. RLIS links tax-lot data to the assessors' databases of three Oregon counties. One of these counties uses a different land use classification than the other two, and of the two that use a similar code, each applies it differently. The result can be amusing. On the regional land use map, a golf course in one county is colored to indicate "commercial" uses, whereas the same use in another county appears as "recreational." For the most part, such anomalies are minimized by the intelligent application of lookup reference tables, to which mapping scripts can refer, as Metro plots the land use maps. Such cross-classification lookup tables are often encountered in researching comparative zoning among different jurisdictions. The most recent release of RLIS Lite, a CD-ROM version of Metro's RLIS in ESRI "shape file" format, offers the land use data in the resulting generalized lookup table version. Thus, the problem has been masked by the reduction of scores of categories to a total of nine. This, however, leads to a loss of precision and to vast variation in land use characteristics among the aggregate classes. For instance, although many SFR (single-family residential) uses are similar, differing mainly in density, vastly different combinations of open space, parking, and building configuration may be found among the various uses all assigned to "COM" (commercial).

Revision of the standard land use coding manual is currently taking place under the sponsorship of the American Planning Association and the Federal Highway Administration. *Land Based Classification Standards* (American Planning Association 1998) is intended to address some of the aforementioned issues. The goal of the sponsors is to broaden the classification to include land cover, environmental constraints, certain characteristics of buildings, and land rights and ownership information. Knowledge of environmental constraints is critical in determining capacity, and buildings occupied by certain uses are unlikely to convert to other less profitable or "lower density" uses. Still, this approach may increase the complexity of a land use taxonomy for monitoring land supply, inasmuch as some of these categories are not normally part of what defines land *use* per se.

Spatiotemporal Representation Models

More than a decade of literature on the development of spatiotemporal data models reveals a somewhat unfinished agenda, with vendor-supplied solutions embedded in commercial software being largely absent or inadequate. A suitable logical implementation of temporally referenced spatial data awaits widespread consensus on the analytical framework within which it is to function.

Three areas of technical advancement will benefit applications in land use, supply, and capacity monitoring:

- The development of conceptual models of temporality for spatial data
- Logical schema for database design and query operations
- Investigation of the availability and quality of spatiotemporal data

Space does not permit a thorough summary of work done in this field, which ranges from conceptual statements of spatiotemporal structure (Langran and Chrisman 1988) to investigations of the utility of temporal extensions to database query languages as developed by Ahn and Snodgrass (1989). Two important implications for parcel-based land information systems emerge from this literature. The first is that, like spatial entities, time and spatiotemporal changes can be modeled as continuous or discrete phenomena. Temporal snapshots of areal patterns can be indexed to moments in time, and spatial overlay processes used to infer changes having occurred between them. This discrete model of time represents states, rather than the events that produce them. Its advantage is that it can be used with historical documents and aggregate change can be computed. For many applications this can be sufficient. However, one disadvantage is that it does not record *when* a particular entity underwent a change. Continuous modeling of temporality, on the other hand, would record events pertaining to spatial objects in the database as they happen, enabling valid spatial queries for any specified time. In the case of the applications discussed in this chapter, such modeling would effectively be implemented with a land use transactional system. Thus, not only could land uses and capacities be measured for any time, but the application, permitting, or inspection events that triggered them could also be uncovered.

Another lesson to be learned from the emerging literature on spatiotemporal databases is that temporal information is not a dimension of spatial objects; rather, it is an aspect of phenomena which can itself be modeled in terms of dimensions. Specifically, it is useful to conceive of a real-world dimension of time (when something happened) and a database time dimension (when something that happened was recorded in the database).

A fully developed model for spatiotemporal land use information will thus integrate the transactional record of land use events with a model of time that can simultaneously reference the existence of other phenomena important to supply or capacity analysis, such as interest rates, housing starts, or projections of population demand. The land use transition component of this model is illustrated in Figure 6.1. In this figure, chronological time is modeled as linear, with land use states (durations) indicated as segments with constant vertical position. Land use states are punctuated by land use events ("indicators"), such as occupancy permits or parcel subdivision, which alter the states. Such an illustration suggests the rudiments of a kind of temporal topology that could be modeled in relational databases similar to its spatial analog. Thus, a given event can be referenced to its prior or succeeding state and a given state can be linked to its beginning or ending event. Furthermore, such a database model would include measurement or verification events

LAND USE TRANSITION

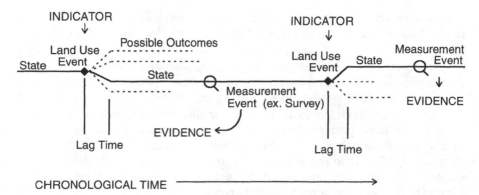

FIGURE 6.1 *Land use transition depicting indicators and evidence*

("evidence"), as well as land use indicators. This would incorporate a means to determine when a given state exists, as well as when knowledge (database reality) of it exists. Making use of these multiple temporal dimensions thus can aid in the analysis of development decisions, for not only could states be queried about given a time, but a check against the transaction time would reveal the evidence that was available at the time a decision was made.

An Analytical Framework

A robust framework for the analysis of spatial, thematic classification, and temporal aspects of land use and supply emerges when we treat these aspects as interrelated but separately modeled phenomena. Sinton (1978) is often credited with introducing an organized framework for temporal and spatial analysis in the GIS literature. He argued that for any observation to be considered useful data, it must be characterized by theme, location (space), and time. His description of map analysis incorporated these aspects (he called them "attributes"). In performing analysis on geographic data, one of these aspects is held constant, a second is permitted to vary in a controlled manner, and the third attribute is measured for its variation with respect to the controlled and fixed aspects. Table 6.2 applies such a framework to a land use database to help organize types of queries regarding land use transition that would be desirable to implement in a plan monitoring system.

CONCLUSIONS

Technical developments in parcel-based land information systems for land use, supply, and capacity analysis will continue to evolve, because new mandates for information spur new methodologies for obtaining it. These methods reveal the inadequacy of existing technology, and the cycle continues when new expectations and revised mandates lead to new methodologies.

Examining existing and recent past practices of land planning with geographic information systems, we can discern methodologies that are proving

TABLE 6.2 Sinton's Analytical Framework Applied to Land Use Data

Structure of Analysis for Land Use Queries			
Example Land Use Query	Constant	Controlled	Measured
What is the areal configuration of a land use?	Time	Theme	Location
What is the land use of a parcel?	Time	Location	Theme
How has the parcel changed configuration?	Theme	Time	Location
When did the parcel change configuration?	Theme	Location	Time
What uses have occupied a parcel over time?	Location	Time	Theme
How long was a parcel engaged in a land use?	Location	Theme	Time

useful but may require technical improvements to data models, query generation, and data acquisition. Continued development in GIS and planning applications is needed for improved information system performance in at least these three areas. In addition, an analytical framework that can query multiaspect databases concerning spatial, thematic, and temporal information about land use and supply needs further refinement in order to take advantage of coming spatiotemporal database functionality.

Urban growth management implies more than providing a supply of convertible vacant land for development. Monitoring the rate of development from raw supply to single-family residential use is only part of the picture. By itself, it fails to provide answers to more pressing questions concerning the success of mixed-use and compact urban design planning objectives. Technical improvements in the management and modeling of land use data are necessary to perform this monitoring. These, in turn, will provide inspiration to implement revised analytical frameworks for examining urban land use, supply, and capacity as dynamic phenomena.

REFERENCES

Ahn, I., and R. Snodgrass (1989). Performance analysis of temporal queries. *Information Sciences* 49: 103–146.

American Planning Association. 1998. *Land based classification standards* ⟨http://www.planning.org/plnginfo/lbcs⟩ (May 19, 1998).

Anderson, J. R., E. E. Hardy, J. T. Roach, and R. E. Witmer. 1976. *A land use and land cover classification system for use with remote sensor data.* Washington, D.C.: United States Geological Survey.

Board, C. 1968. Land use surveys: Principles and practice. In *Land use and resources: Studies in applied geography.* Institute of British Geographers Special Publication No. 1, 29–41.

Bollens, Scott A., and David R. Godschalk. 1987. Tracking land supply for growth management. *Journal of the American Planning Association* 3: 315–327.

Boyle, A. R. 1987. The user profile for digital cartographic data. *Procedure of Auto-Carto 8.* Vol. 1. Baltimore Md.: ACSM/ASPRS, 12–15.

Chapin, F. Stuart, and Edward J. Kaiser. 1985. *Urban land use planning.* Urbana: University of Illinois Press.

Clawson, Marion, and Charles L. Stewart. 1965. *Land use information: A critical survey of United States statistics including possibilities for greater uniformity.* Baltimore, Md.: Resources for the Future.

Couclelis, H. 1991. Requirements for planning-relevant GIS: A spatial perspective. *Papers in Regional Science* 1: 9–19.

Craig, W. 1980. "Monitoring Minnesota's land using public records." Ph.D. diss., University of Minnesota at Minneapolis.

De Grove, John M. 1992. *Planning and growth management in the states.* Cambridge, Mass.: Lincoln Institute of Land Policy.

Easley, V. Gail. 1992. *Staying inside the lines: Urban growth boundaries.* Planning Advisory Service Report No. 440. Chicago: American Planning Association.

Godschalk, David R., Scott A. Bollens, John S. Hekman, and Mike E. Miles. 1986. *Land supply monitoring: A guide for improving public and private urban development decisions.* Boston: Oelgeschlager, Gunn & Hain, in association with the Lincoln Institute of Land Policy.

Hudson, G. D. 1936. The unit area method of land classification. *Annals of the Association of American Geographers* 2: 99–112.

Jordan, T. 1982. Division of the land. In *This remarkable continent: An atlas of United States and Canadian society and cultures.* J. Rooney, W. Zelinsky, and D. Louder, eds. College Station: Texas A&M University Press, 30–46.

Knaap, Gerrit, and Arthur C. Nelson. 1992. *The regulated landscape: Lessons on state land use planning from Oregon.* Cambridge, Mass.: Lincoln Institute of Land Policy.

Langran, G., and N. Chrisman. 1988. A framework for temporal geographic information. *Cartographica* 3: 1–14.

Logsdon, Miles G., Earl J. Bell, and Frank V. Westerlund. 1996. Probability mapping of land use change: A GIS interface for visualizing transition probabilities. *Computers, Environment, and Urban Systems* 6: 389–398.

McLaughlin, J. D. 1984. The MPC concept: Current status, future prospects. In *Seminar on the multipurpose cadastre: Modernizing LIS in North America.* Bernard Niemann, ed. Madison: University of Wisconsin Institute for Environmental Studies.

Mildner, Gerard C. S., Kenneth J. Dueker, and Anthony M. Rufolo. 1996. *Impact of the urban growth boundary on metropolitan housing markets.* Portland, Oreg.: Portland State University.

Mitchell, W. B., S. C. Guptill, E. A. Anderson, R. G. Foges, and C. A. Hallam. 1977. *GIRAS: A geographic information retrieval and analysis system for handling land use and land cover data.* Professional Paper 1059. Reston, VA: U.S. Geological Survey.

National Research Council (NRC). 1980. *Need for a multipurpose cadastre.* Washington, D.C.: National Academy Press.

———. 1983. *Procedures and standards for a multipurpose cadastre.* Washington, D.C.: National Academy Press.

Portland Metro. 1997. *Performance measures: Draft recommendations from the executive officer to the Metro Council concerning the urban growth management functional plan.* Internal memo.

Sinton, D. 1978. The inherent structure of information as a constraint to analysis: Mapped thematic data as a case study. *Procedures of first international advanced study symposium on topological data structures for geographic information systems.* Vol. 7. Cambridge, Mass.: Harvard University Graduate School of Design, Laboratory for Computer Graphics and Spatial Analysis.

Stamp, D. L. 1940. Fertility, productivity, and classification of land in Britain. *Cartographic Journal* 96: 389–413.

Stein, Jay M., ed. 1993. *Growth management: The planning challenge of the 1990s.* Newbury Park, Cal.: Sage Publications.

Urban Renewal Administration, Bureau of Public Roads. 1965. *Standard land use coding manual.* Washington, D.C.: United States Government Printing Office.

Vrana, Ric. 1997. "Monitoring urban land use transition with geographic information systems." Ph.D. diss., University of Washington.

Vrana, Ric, and Kenneth J. Dueker. 1996. *LUCAM: Tracking land use compliance and monitoring at Portland Metro.* Report to Metro, Portland, Oregon. Portland: Center for Urban Studies, Portland State University.

COMMENTARY: GIS AND REMOTE SENSING IN A GROWTH MANAGEMENT CONTEXT

Frank Westerlund

My comments underscore Ric Vrana's general observation that a gap exists between growth-management related monitoring approaches that have emerged in theory and, in some places, in practice, and the level of application of GIS tools that is typical today. This gap relates, in part, to the differing limitations posed by the parcel as a unit of data analysis, or as a unit of data collection. For many purposes of areawide analysis, the parcel is an inappropriate unit (see Chapter 3). Yet, as shown in the Portland Metro work, parcel data can be aggregated to meaningful larger analytical units. Vrana's table showing "status" and "rate-of-change" variables for monitoring plan "performance" also lists several types of aggregated units appropriate to different variables, many or most of them derived from parcel data.

We are also limited in various ways by data collection at the parcel level. Even where aggregation of parcel data may be workable, it may not be the most efficient approach. Study of demand-side factors, such as infrastructure capacity and environmental system constraints, certainly requires getting outside a parcel framework. At the same time, looking at how individual parcels respond to these forces in terms of land availability, price, likelihood of conversion, development cost, and so forth, may be useful.

For the supply side of land monitoring, there are practical reasons for making significant use of parcel-level data, as illustrated by the techniques used by Portland Metro. The estimation of where development is likely to occur was admittedly less successful in its use of such parcel data as building-to-land-value ratios. This is not so much a criticism of using this particular measure, as a suggestion that building-to-land-value ratios should be regarded as only one of many supply-side indicators. These ratios cannot substitute for demand-side data or consumer preference data of the type developed in the demand simulation approach described in Chapter 8.

Vrana briefly mentions the technical problems associated with what he terms differentiating between multiple-use parcels and mixed-use areas. Trying to capture the nature of residential and commercial integration occurring within parcels is one of a number of supply-side questions that may require subparcel levels of data collection, in addition to nonparcel coverages such as the environmental data typically overlaid on parcels. The presence of and some of the characteristics of multiple uses (e.g., number of dwelling units and square feet of retail use) can be readily identified within a parcel database to relate these uses to performance or functional design standards. However, this may imply the need to capture spatial configurations of multiple uses within parcels.

Capacity analysis may also require going to a subparcel level anyway, to fully analyze the remaining development potential of partially developed parcels out to the envelope established by zoning or other regulations. Perhaps this can be done crudely by looking at the floor area ratio (FAR) of existing development as compared with the maximum FAR, if set by zoning. However, considering that the typical zoning ordinance today may have many pages of development standards for even a single category, such as multifamily (including variable setbacks and building heights, modulation, and other spatial controls for structures within a parcel), and that some of this information probably should figure into capacity analysis, we need to be thinking about capturing the subparcel spatial configuration of existing development.

I heartily agree with Vrana's three categories of needed development in GIS technical methods, encompassing land use classification, legacy data, and spatiotemporal modeling issues.

Land use classification taxonomies constitute an area that needs a thorough revisiting. In large part, our planning policies and regulations are based on land use classification, yet this subject has seen little development. The American Planning Association's current project on such classification is timely. Possibly no single typology can be fully adequate for land use description. Needed are the physical dimensions represented by land cover and structure type and form, and the economic and human activity dimensions. Traditional systems, those such as included in the *Standard Land Use Coding Manual*, confused these dimensions and, as Vrana points out, traditional hierarchical structures often do not make sense in terms of land use functionality and impacts. Spatial hierarchies that relate to contemporary ways of viewing development may work better—such as, for example, morphologically defined general categories (e.g., "commercial strip") combined with subcategories for the kinds of uses found within them.

Moreover, Vrana's model of spatiotemporal representation is compelling in its combination of a "states" approach and an "events" approach. As a major aspect of growth management, monitoring concurrency requires comparison between two spatiotemporal streams: the rate and location of permitted (and actual) development, and the rate and location of infrastructure improvements.

Capturing and using legacy data may be more of an institutional problem than a technical one. The failure of the transactional record to fully reveal actual change, or the lack of it, remains a problem only partly caused by parcel dynamics. Some permitted development never occurs, or occurs contrary to permission. This is where an independent monitoring tool such as remote sensing can be useful. In operational terms, a remote sensing approach to land monitoring means acquiring a periodic time series of digital image data at appropriate intervals and of sufficiently high resolution to discern development or development in progress, including outlines of individual building structures. Images can be viewed as a backdrop to a parcel coverage for routine, systematic checking against the associated parcel transactional database. Recent advances in high-resolution digital imaging from both aircraft and satellites offer potential for delivering such imaging according to a programmed schedule and at reasonable cost.

However, much information of a physical nature is essential to a monitoring process. Today's remote sensing technology has almost unlimited spatial and temporal flexibility, and it is parcel independent. Derived information is easily aggregated to parcels or other units. Subparcel detail can be attained at fine levels of spatial resolution, even with satellite systems capable of near-continuous monitoring in time.

A number of commercial satellites are now being launched with these capabilities. One of them, the Earthwatch Quickbird satellite, will produce 0.8 meter resolution multispectral imagery using a pointable sensor that looks forward and backward within the satellite's track to provide stereo views and digital elevation models with vertical resolution in the 1-to-2-meter range. This technology offers the potential ability to produce building footprints, building heights, and mass configurations, as well as subparcel impervious surface and natural cover, and to do this frequently and (it is hoped) at reasonable cost. Quickbird will produce up to five snapshots per day over any location, with the data distributed over the Internet. Such data collection ability approximates continuous monitoring and begins to blur the distinction between events and states.

The possibilities here are many. However, much work remains to be done to make remote sensing an operational component of a monitoring system. Particularly needed are developments in the image analysis processes to substantially automate information capture. Because remote sensing depicts only physical features and characteristics of the earth's surface, human-made and natural, anything related to human activity or socioeconomic conditions is an inference from physical features. Even to provide meaningful physical description requires interpretation, by human or machine, of raw image data. For this reason, remote sensing can be only a part of a larger set of methods for growth management monitoring. One promising avenue to replacing labor-intensive interpretive tasks by automated analyses methods is the application of algorithms for spatial pattern recognition, which have been developed for military remote sensing applications and are now beginning to be applied to civilian sector uses.

COMMENTARY: FUTURE CONSIDERATIONS—DATA, TECHNOLOGICAL ADVANCES, AND MODELING

Marina Alberti

The chapter covers a lot of territory and raises a number of challenging methodological and technical questions. I will focus only on a few points, based on my experience with parcel-based GIS as used for environmental planning and management.

Conceptual Framework

Three assumptions are made: (1) Technical means influence the methodology for managing information, (2) GIS has provided the tools to make spatial data more available, and (3) increased availability and access to spatial analysis tools have increased the efficiency of existing mapping and suggested new ways of monitoring the development and supply of land. I share this evolutionary view. Often technical solutions result from a vision about the future. This suggests that we need to ask not only what information is relevant to solve our current questions, but also what land use, supply, and capacity information we think will become relevant for planners 10 to 15 years from now. For example, when environmental planners ask public agencies for more disaggregated data related to human-induced environmental stresses, the typical answer is that this information is not even collected. I refer in particular to parcel-level data on the consumer behavior of businesses and households, which is relevant to monitoring and managing land. As planners we have responsibility to identify the information and monitoring capacity we should make available as well as the technical implications of making it available.

Because building databases does not happen overnight, the question is whether we can identify the data requirements and formats we need over the long term. Time is an important resource. How can we spend it efficiently?

In response to this question, data and information must be treated differently. Data are observations about a certain phenomenon. Information is data that have been processed and put into a context. With respect to "maintaining a reserve land supply," for example, this chapter stresses the difference between monitoring where and when development or redevelopment is occurring and formulating an estimate of where it is likely to occur. I suggest that the "indicators" of land supply and capacity may become obsolete as, over time, we change the definitions governing environmental constraints on development or as land markets change. Although the land use record may still be appropriate to describe a given use over the long term, such indicators may become obsolete because of changes in policies, specific targets, and methodologies. This implies that we must distinguish between the data we collect to describe land use and the indicators selected to predict change or assess performance.

Technological Development

A second point refers to technological advances and the distinction between methodological and technological problems. Burrough (1986) suggests that sometimes problems perceived as conceptual are indeed technological problems. For example, consider two technological advances that in the last two decades have changed the way we think about potential solutions and methodologies: raster and vector GIS, and on-line GIS.

Burrough (1986) explores the first example in his discussion of the choice between raster and vector data structure. Raster and vector data structures are two different approaches to modeling geographic data. Until a few years ago, Burrough says, it was conventional wisdom that these data structures were irreconcilable alternatives, because raster systems required substantial computer memory to store and process data at the same spatial resolution as that which could be obtained with vector data structures. Yet as compared with raster systems, vector systems could not handle as easily certain kinds of data manipulation, such as polygon intersection, spatial averaging, or neighborhood operations. We were then faced with the choice of using vector structures with limited spatial analysis potential or raster structures requiring large computer storage capacity. Burrough argues that this example illustrates a technological problem. Indeed, many of the algorithms that had been developed for vector data structures had a raster alternative. Moreover, the problem of storage volumes required for raster data could be reduced by such means as compact raster structures (e.g., run length codes or quadtrees). Now that the technological issue has been resolved, the nature of the problem is changed. The problem is not a necessary choice between raster and vector formats, inasmuch as both are valid methods for representing spatial data. Rather, it is how to use both in an integrated way.

The possibility of accessing GIS data on-line is another technical solution that will change the nature of some of the problems we may now consider relevant. So far, GIS systems and public access to georeferenced information have been treated as two separate issues. Until a few years ago, producing and storing high-resolution maps and making them available to the public were considered irreconcilable tasks. Developments in on-line GIS have demonstrated that the two are not only reconcilable but that they can actually help each other. For example, high-resolution maps will provide additional information, and increased access will enhance data quality control. At this point we do not know what additional problems and possibilities this will create.

Environmental Modeling

My third suggestion is to consider incorporating ecological modeling in the design of land monitoring systems. Addressing the interface between land use and environmental planning, and linking land and environmental monitoring, require sharing data across a broad range of agencies, each with different tasks and needs. In doing so, several methodological issues arise, which include the spatial units considered, the land use taxonomies used, the spatiotemporal representation selected, and dealing with uncertainty.

The first issue involves two levels of spatiality. One is the spatial unit of analysis. Is the parcel a good way of representing human stressors on the environment? As suggested in chapter 3, parcel boundaries may be too dynamic, and the parcels themselves may be too heterogeneous for many attributes of interest and for their effective use as the sole unit of land monitoring. The "mixed spatial units" approach will be appropriate for many types of environmental modeling. A second related aspect is the boundary of the inventory. When relying on land records from local jurisdictions, the environmental analyst must beware of the technological and institutional implications of combining administrative and ecological boundaries.

Land use and land cover are often used interchangeably, but are indeed two distinct representations of the landscape. Because both classification systems are used frequently in land use and environmental planning, it is critical to understand their relationship. Land use and land cover categories can overlap or be nested within each other in ways that affect their usefulness as input data for biophysical models. The question is whether land use is a good predictor of land cover. Is there a way to compare land use with land cover? The problem is that some land uses (e.g., residential) may have different land cover configurations, and areas with homogeneous land cover (e.g., forested land) may have different uses within them. Both have implications for assessing biophysical impacts (on, for example, water runoff). A multicoded approach can best serve the purposes of representing land.

An important aspect to consider for spatiotemporal representation in environmental modeling is scale. Spatial metrics are scale-dependent or are relevant to environmental processes operating only at specific spatial scales. Moreover, environmental impacts have different effects and implications in the short, medium, and long term. We need to explore ways in which events happening at different spatial and temporal scales can be recorded and represented.

Finally, surprise events such as fires, floods, earthquakes, and economic cycles are unpredictable but leave important marks on land use. Further research is necessary to address the methodological and technical complexities involved in incorporating these aspects in a land monitoring system.

These are some of the most important methodological and technical challenges raised by Vrana that are related to linkages between land and environmental capacity monitoring, and they will have to be addressed in the near future.

References

Burrough, P. A. 1986. *Principles of geographical information systems for land resources assessment.* Oxford: Clarendon Press.

Urban Renewal Administration, Bureau of Public Roads. 1965. *Standard land use coding manual.* Washington, D.C.: United States Government Printing Office.

Vrana, Ric. 1997. "Monitoring urban land use transition with geographic information systems." Ph.D. diss., University of Washington.

Vrana, Ric, and Kenneth J. Dueker. 1996. *LUCAM: Tracking land use compliance and monitoring at Portland Metro.* Report to Metro, Portland, Oregon. Portland: Center for Urban Studies, Portland State University.

7

Data Sharing and Organizational Issues

Anne Vernez Moudon and Michael Hubner

Organizations are related dynamically to the technologies that serve them—they influence each other as they change over time. This relationship is complex and evolves slowly. In the arena of local government, organizational change often trails behind technological change and thus slows the creation of conditions in which geographic information systems (GIS) can be used effectively for land supply and capacity monitoring (LSCM). As a highly data-dependent activity, LSCM stands to benefit from recent and ongoing advances in land information technology (especially GIS and GIS data creation), but is equally apt to be handicapped by institutional resistance to the altered practices implied by these advances.

Land monitoring relies on many data sources to meet increasingly complex and broad-ranging requirements. Most typically carried out by planning departments, LSCM utilizes both in-house data (e.g., land use, zoning, planned land use) and data produced by other agencies and departments (e.g., assessors' data, utilities infrastructure). The growing complexity of data collection, handling, and analysis has required extensive commitments in skilled staffing, equipment, and software purchase and development—often well beyond that necessary for traditional planning activities. The magnitude of technical support required partly explains why LSCM is not more widespread: It exceeds

Monitoring Land Supply with Geographic Information Systems, edited by Anne Vernez Moudon and Michael Hubner ISBN 0 471371673 © 2000 John Wiley & Sons, Inc.

the resources—both human and financial—available for planning in most jurisdictions.

Two recent developments in land information systems will likely affect the evolving practice of land supply monitoring: the emergence of parcel-based geographic information systems (PBGIS) as a standard element of the information infrastructure of local governments,[1] and the adoption of GIS by a wide range of departments and agencies within local government (see Chapter 1). The latter development is the subject of this chapter, with emphasis on the opportunities that GIS applications (particularly PBGIS) in local government present to improve data availability for LSCM.

This chapter reviews current thinking about GIS database coordination within the organizational structures of local government and explores the barriers and opportunities for improving data sharing for LSCM. It draws from several sources, including the literature on planning applications of GIS, the survey of land monitoring agencies carried out at the inception of the research undertaken for this book, and the presentations and discussions of the May 1998 conference session dedicated to institutional and organizational issues, which was presided over by Nancy Tosta.

GIS IN LOCAL GOVERNMENT

The current boom in GIS applications within local government suggests that implementing LSCM need not rely on an independent database development effort. Building a land monitoring database can be largely a matter of coordinating the work of different departments to obtain requisite data in a workable format and at an adequate level of accuracy and timeliness for planning purposes. PBGIS increases the opportunity for utilizing data generated by the various departments, much of which are collected and maintained at the parcel level. However, for LSCM to successfully leverage these data, solid organizational structures for data sharing must exist. Not exclusive to the topic of LSCM, data sharing issues are discussed in a varied body of literature that will be summarized in this chapter. To lay the groundwork for discussing data sharing, the following section reviews the range of GIS organizational structures within which land monitoring occurs.

GIS Organizational Structures

GIS organizational structures pertain to both the institutional and the computing environments supporting the technology. Nedovic-Budic and Pinto (1998, 16) state that "GIS implementation in organizational settings is . . . a

---

[1] As a GIS that includes ownership parcels among its primary spatial layers, PBGIS add to older nonspatial land information systems by allowing for the storage, retrieval, visualization, and analysis of parcel data based on its spatial dimensions (including location, size, adjacencies, "neighborhood" context). Hence, patterns of land with given attributes and characteristics can be identified to further define and delineate land supply and capacity within urban areas.

complex process that involves installing, maintaining, and using a system in [institutional] environments that have diverse functions, tasks, resources, motifs, interests, and goals." Such is the case today within local government, where significant resources are spent in establishing and coordinating GIS activities.

The formalization of a basis for GIS coordination and data sharing addresses local governments' "need for accurate, current, and applicable data—and the need to reduce data redundancies, gaps, and costs" among departments and agencies (Kollin et al. 1998, 30). The important dimensions of GIS organizational structure involve the levels at which coordination occurs, the loci of control and standardization, and the relationships that determine the flow of data to and from GIS users and data providers. These dimensions can be assessed succinctly within four basic models for local government GIS (Public Technology Inc. et al. 1991). Each model implies different levels of authority and degrees of accessibility to and exchange of GIS data.

- *Single-department GIS.* Located within, as well as controlled and driven by, the needs of one agency, single-purpose GISs are usually not associated with significant amounts of data sharing and coordination. Control is located within departments, and the flow of information is circumscribed by departmental boundaries.
- *Shared GIS dominated by a single department.* Individual departments (typically engineering, public works, or planning departments) commonly dominate GIS data-sharing relationships. Although formalized arrangements for sharing data can be made between the dominant department and other departments, exchange is limited by the degree to which the data collection, representation, and management practices of the host department are compatible with those of other departments.
- *Multidepartmental GIS.* Data sharing and coordination may be formalized among multiple departments in partnership with each other. Such a GIS is typically managed by a separate office or department and guided by an interdepartmental body. This model entails the development of "one set of strategic components that can be used to develop diverse applications" (Public Technology Inc. et al. 1991, 29). Its success depends on the tenor of broad organizational relationships and the degree of general coordination that already exists between departments (especially within large, complex bureaucracies).
- *Multiagency GIS.* Finally, GIS may be coordinated among multiple agencies and jurisdictions (e.g., cities, counties, and metropolitan planning organizations). Interjurisdictional differences in policy and data management will compound the interdepartmental differences that may already exist. This is especially true at the regional scale, where more than one county is typically involved. Interdepartmental coordination within jurisdictions will likely have to be worked out *before* this model can be successfully implemented.

Of equal importance to organizational structure is the structure of the computing environment, which at once mirrors and shapes the organizational structures. Bollens and Godschalk (1987) described three primary computing system options for an "Automated Land Information System" that continue to
have relevance to land monitoring practice:

- *Centralized.* All users are tied to one centralized processing facility or
  department. Although this model is archaic by late 1990s standards, many
  successful local government information systems used for LSCM still rely
  on older mainframe computers.
- *Decentralized.* This type of system comprises multiple independent processing centers serving the needs of separate departments, or clusters of
  departments. Despite widespread coordination efforts, many local governments' GISs continue to exist as multiple independent installations.
- *Distributed.* In this model, a central library of geographic information
  can be downloaded to different departments. Update responsibilities are
  distributed to the departments that most directly deal with each data
  type. A distributed system provides for increased responsiveness and
  flexibility of use for all participants and has been greatly facilitated by
  advances in computer networking technology. It is the model of choice
  among newly implemented GISs.

The hierarchy of GIS organization—from single department to multijuris-
diction, from mainframe to distributed network—suggests an evolutionary
development from disorganized, piecemeal adoption to fully coordinated and
integrated systems. In the real world of practice, however, the evolution of
GIS is a slow and incremental process. Recent survey findings illustrate this
reality. Kollin et al. (1998) found a majority of jurisdictions having neither a
single focal point for GIS coordination nor a clearinghouse for making GIS
data accessible to multiple departments. Our own survey found, as well, that
it is common for many agencies within a jurisdiction to have separate GIS
installations, characterized by redundant and incompatible data collection and
maintenance practices (see Appendix A). The persistence of GIS fragmentation within local government stems from both sizable investments in current
(fragmented) systems and the quasi-independence of many local government
functions. These conditions suggest that the current state of affairs is likely
to continue until the benefits of exchange and interaction between the separate
departments that manage GIS databases are perceived to exceed the costs of
altering the organizational status quo.

Sharing GIS Data

Shared GIS within local government is characterized as being (1) *multipurpose,*
serving many different functions, such as tax assessment, capital facilities

management, and planning, and (2) *multiparticipant,* involving a number of different departments and staff.

Sharing GIS data is a central concern for jurisdictions nationwide, yet it is difficult to achieve in practice (Charlton and Ellis 1991). Examining the state of GIS practice, Nedovic-Budic and Pinto (1998) reviewed the literature as background to intensive case studies of GIS data sharing in five representative jurisdictions. They confirm the primary reasons for sharing to be greater data consistency, enhanced organizational efficiency (hence, reduced costs), and reduced redundancy of database development. Such benefits also help small departments to adopt technology that they could not afford independently. By expanding the user group for GIS, data sharing is also often the only way that jurisdictions (including large ones) can justify the three-to-five-year length of time necessary before new systems are fully operational. Similarly, shared systems are likely to reduce the costs of implementing a comprehensive land supply monitoring program, even within large jurisdictions.

The same authors assert that programs to implement GIS data sharing must acknowledge and respond to the different purposes and uses of geographic data. They must establish each participant's level of authority, proprietary interest, level of access, liability, and responsibility. In addition, different levels of skill and technical familiarity will have to be negotiated (although this problem should be partially be mitigated over time by increased GIS training among local government professionals, especially planners). Successful implementation will require support and leadership at high levels of local administration. As implementation proceeds, demonstrated progress (particularly delivered products) will help ensure continued political support.

It is important that long-term coordination also be considered. GISs are often built incrementally, department by department. For example, Seattle and several of its suburbs required approximately a decade of GIS development before achieving operationality for policy support. Within the context of long-term implementation, GIS coordination must address the potential for future change, both technological and within and between organizations.

Furthermore, the establishment of *standards* has been hailed as a central element necessary for successful integration of GIS resources. Several national efforts have been under way to develop and foster the adoption of GIS standards. The Federal Geographic Data Committee (FGDC) has drafted standards for both spatial and attribute data, including cadastral standards (von Meyer and Moyer 1994; Federal Geographic Data Committee 1996). The Open GIS Consortium, Inc., is an industry-led effort to foster the sharing of GIS data, particularly across platforms in networked computer environments (Heikkila 1998). Standards are "an important means for bridging communication gaps . . . [but] the effectiveness of any standard depends upon the value of the standard as perceived by a sufficient number of people who are willing to observe it" (Fernandez-Falcon et al. 1993, 22).

The importance of standards applies to both spatial data and metadata, as well as to the use of GIS data. "User design standards" are intended to "influence the design and implementation of GIS including attribute data schemes, coding rules, map accuracy, quality control, and map design" (Fernandez-Falcon et al. 1993, 22). Cost-conscious state and local government agencies increasingly perceive such standards as necessary. This necessity extends to the practice of land monitoring, in which data of widely different degrees of accuracy and scales (for example, wetlands versus tax lots) may be combined to assess the suitability of land for future development.

Establishing useful common standards (from the national to the local level) is a slow and difficult process, involving input from a wide range of users as well as piecemeal adoption across institutions and organizations. The local, fragmented basis of planning and land use management in the United States exacerbates the difficulties in enforcing standards for GIS data, particularly land records. The most effective approach may be for regional bodies to establish standards and then provide incentives for local jurisdictions to migrate toward adopting those standards over time.

PARTICIPANTS IN DATA SHARING FOR LSCM

In order to address successfully the issues in GIS data sharing most germane to land supply monitoring, one must recognize the various agencies that could and should be involved in data sharing for LSCM. The nature of these agencies' activities and their specific roles within the land use change and land monitoring processes directly affect the types of data they generate and use, and the way in which they store and retrieve these data.

This section of the chapter reviews the principal players in the local arena, who are commonly involved in managing or defining the juridical, fiscal, and use (existing and potential) characteristics of urban land. Not only are these players numerous, but the scope and nature of their activities vary greatly: (1) they comprise both public- and private-sector interests, (2) they perform different functions, with implications for specific data requirements and units of measurement, (3) they operate within distinct time frames and degrees of regularity or continuity in their activities, all of which affect the data they collect and maintain, and (4) they are all associated with specific interests in access to detailed and accurate information on aspects of land supply and capacity.

Public Sector

The following six functional areas must be considered in the public-sector arena.

Advance Planning The institutional context of land monitoring highlights the distinction between "advance planning" and "development management" (Kaiser et al. 1995) (also termed "higher-order" planning and "lower-order" planning by Birkin et al. [1996]). Advance planning, which involves strategic long-term considerations (encompassing both comprehensive planning and regional forecasting) uses land data primarily for analytical and evaluative purposes. Development management, on the other hand, deals with the everyday concerns of plan implementation (such as zoning and subdivision review) and generates land data primarily within the realms of site- or project-specific record keeping.

Land supply monitoring has thus far primarily occurred within the province of advance planning, which usually takes place in cycles of 5 years or more and generally addresses land supply over long time frames (20 years or more). Land monitoring for advance planning is often episodic or sporadic in occurrence. It has typically been concerned with zonal geographies (e.g., planning areas, zoning districts, areas of contiguous land use), but may also include parcel data, depending on data availability and the size of the jurisdiction.

As the local government function most involved with monitoring supply and capacity, advance planning is potentially interested in a full range of land information and, hence, often functions as a focal point for coordinated data sharing. Advance planning also generates its own data—including data on existing land uses, planned land uses, long-term plans for infrastructure, major policies affecting development, and allocations of regional demand forecasts.

Development Management   The development management function oversees land use change and the improvements to land effected through the development process (Kaiser et al. 1995). It involves the creation of regulations and the administration of laws and programs that implement the policies established by advance planning. Development management covers permitting (e.g., building, occupancy), subdivision and site design approval, zoning (including adjustments over time, such as variances and rezones), and the administration of a variety of programs shaping the ongoing development of urban land (e.g., transfer of development rights, adequate public facilities ordinances, tax incentives).

Development management generates large amounts of data relevant to LSCM, such as records of public actions concomitant to applications for and approvals of private development, and requirements of specific regulations affecting land and its development. Insofar as it is the primary function involved in plan implementation, development management can benefit from LSCM to provide feedback on the effectiveness of that implementation, including detailed information on the location, mix, and density of new development within land use areas and zoning districts. In addition, this local government function benefits from timely information on the short-term land supply to respond appropriately to private-sector priorities.

Land Records Management and Assessment   Land records and assessment functions focus primarily on the juridical and fiscal components of land, and only secondarily on its physical and use characteristics. Land records management is an ongoing activity, with property description updates and ownership transfers occurring on a continuous, often daily, basis. The maintenance of property records overlaps considerably with assessment for local property taxation. The assessor's land records contain data for tax lots on existing land use, physical descriptions of land and improvements, valuation (for land and improvements), ownership, rights in property, tax status (including delinquent lands), and other data relevant to the assessment process.

The dependence of land supply monitoring on the assessment function has been and will continue to be crucial, because assessors' files are a primary source of parcel-level data. The relationship between monitoring and the assessment function is becoming more important with the adoption of PBGIS. Conversely, property assessment for taxation has a direct interest in estimates of the future developability of land as a factor for appraising land and improvement values. As the assessment function becomes increasingly automated and reliant on GIS as an analytical tool, accurate and detailed data on the existing and future development capacity of land can and likely should help to make the assessment process both reliable and transparent under public scrutiny. For these reasons, the importance of building and maintaining databases on land supply and capacity that are useful to the assessment function will likely grow.

Economic and Community Development and Housing

Local agencies devoted to economic and community development and housing focus on the welfare of households, businesses, and distressed neighborhoods. With geographic areas of concern ranging from the neighborhood to the region, data collection by these agencies concentrates more on activities than on land or land use. This includes data on firms, employment, trade and intraurban movement of goods, different types of economic activity, housing conditions and costs, and indicators of housing needs.

Analyses of economic and housing conditions have typically been unconcerned with detailed considerations of the spatial aspects of land supply. They have only indirectly relied on GIS, particularly PBGIS, and they have not overlapped significantly with LSCM. In the current growth management context, however, interests may be converging to provide incentives for data sharing. In the area of housing, for example, affordability is increasingly linked to land supply and the impacts of land regulations on specific locations and sectors of the housing market. LSCM itself is beginning to extend, in explicit ways, into the realm of monitoring housing stocks and prices. Similarly, the impact of land supply on employment-generating activities is becoming important to these functions, in that land use restrictions shape the location of future economic activity and employment growth. In this area, monitoring efforts are beginning to assess the adequacy of the land supply for future business expansion. Finally, land supply is a significant issue for distressed districts, particularly in older central cities, where the development potential of industrial brownfields, vacant lands, and tax-delinquent properties is a growing focus of community development efforts.

Environmental Planning and Protection Environmental agencies deal with the interactions between urban activities, urban development, natural hazards, and general environmental quality. Operating at many levels of government, they include county and municipal departments (e.g., water resource management), state agencies (e.g., departments of natural resources), and federal agencies (e.g., Federal Emergency Management Agency, Environmental Protection Agency).

The data generated by these agencies have been crucial for identifying the buildable land supply. They include topographical and soil conditions, floodplains, wetlands, geologic hazards, resource lands and natural habitats, and other areas subject to special restrictions.

Conversely, urban growth—comprising both new development at the fringe and intensification of use in areas already developed—is an ongoing environmental concern. Environmental agencies were early users of GIS, with one of their basic activities being to monitor human impacts on the environment. Environmental impact monitoring can greatly benefit from accurate and disaggregated data on the future developability of land within urbanizing regions. As increasing attention is paid to the cumulative impacts of small-scale land use changes, high-resolution GIS data (particularly parcel-level) can help to quantify and to gauge the importance of urbanization impacts in environmentally sensitive areas (e.g., watersheds, wetlands, shorelines, and riparian corridors) or in areas at risk for natural hazards (e.g., from fire, flood, and landslides). Such detailed information is needed to design environmentally appropriate regulations and to prevent or mitigate the risk of property damage from hazardous conditions.

Another expanding area of overlap between land supply monitoring and environmental planning lies in the strengthened federal mandates to coordinate land use and transportation planning to improve air quality (see Chapter 8 and Alberti's "Commentary" to Chapter 6 for further discussion).

Public Service Provision The provision of local public services is an ongoing activity with significant impacts on land use. The maintenance and expansion of major public infrastructure (water, sewer, and transportation) directly services land for existing and future development. Other public services, such as education, recreation, and public safety, have traditionally been considered to have a less direct, yet important and long-term, impact on the location and intensity of urbanization.

As discussed in previous chapters, levels of servicing are primary considerations in evaluating the capacity of the land supply. Public service providers generate data used to make these determinations and rely on capital facilities plans to establish levels of service for planned infrastructure elements. Geographic units representing service demand data are typically zonal (e.g., Forecast Analysis Zones [FAZs], subareas, or districts), with service capacities represented on a zonal, or, more commonly, a network basis (especially transportation, water, and sewer).

Structured to perform ongoing daily functions, most public service provision departments depend on institutional planning in the medium term, through two- to seven-year capital budgeting. However, large-scale capital improvements, such as limited access highways, bridges, reservoirs, primary water lines, trunk sewer lines, and sewage treatment facilities, are planned over long time horizons of up to several decades.

Land supply and capacity information is essential to public service planning and provision because it enhances the ability to estimate future service needs in growing areas (including urban expansion and infill or redevelopment). It also allows accurate calculations of the costs and benefits of extending services, thereby, for example, facilitating annexation studies.

Private Sector

The private sector has not traditionally been included as a data-sharing partner in local government GIS. However, the utility of incorporating market-oriented data, much of which is generated by the private sector, has long been emphasized by LSCM practitioners and in the literature on land monitoring (Godschalk et al. 1986). Over time, intensified efforts to engage the private sector in the design of land monitoring systems may lead to formalized data sharing arrangements (see Rolfe's "Commentary" to Chapter 3). These data may include recent sales records, properties currently or recently on the market, derived, for instance, from Multiple Listing Service data, and information on the development feasibility of different types of projects in different locations.

Private-sector activity can be divided into two categories—real estate transactions (buying and selling) and land development or redevelopment. Real estate transaction functions are performed continuously on a daily basis. Transactions are short term (generally ranging from days to months) and typically concerned with the parcel of ownership as the unit of land considered, meaning tax lots, small aggregations of tax lots or, in the case of condominiums, dwelling units. Land development and redevelopment activities also occur on an ongoing daily basis. Land development is typically performed in the short to medium term, ranging from months to several years. Units of land considered range from the ownership parcel to aggregations of parcels. In addition, private development processes are subject to significant surges, which periodically put pressure on public-sector functions.

Primary private-sector activities related to land development include site search and suitability analyses (e.g., finding an appropriate site for a given land use), individual project design and development feasibility analyses (e.g., finding an appropriate land use for a given site), marketing development products, securing public approval for projects, preparing and servicing sites, and land subdivision or assembly.

The private sector has successfully made limited use of GIS databases to support development strategies, particularly for evaluating potential office and retail locations (Castle 1993, 1997; Thrall 1997). It has also relied on, and indeed demanded, public-sector information to assess periodically the impacts of land policy and regulations on its activities.

Private-sector interests are likely to become further invested in LSCM for several reasons. First, they can benefit from information on the development potential of specific areas or sites within a region. Second, LSCM can help secure a predictable development climate (as illustrated in the case of Montgomery County, Maryland; see Chapter 4). Further, because LSCM can provide a basis for balancing land supply and demand over the long term, it can help to stabilize regulatory frameworks and thereby inject increased predictability in the future of development markets. The availability of publicly accessible, credible databases on land reduces the risks associated with land development in a competitive and dynamic environment (Dueker and DeLacey 1990).

ISSUES IN DATA SHARING FOR LSCM

By crossing departmental and jurisdictional lines, and by requiring considerable data integration, LSCM potentially increases the already problematic nature of GIS data sharing. Specifically, adapting local government GIS for regional monitoring raises several issues. First, it is necessary, but also problematic, to reconcile local government's primarily administrative purpose in acquiring and processing data with the policy orientation of LSCM. Second, achieving multijurisdictional coordination entails complex organizational structures. Third, securing access to market-oriented and privately generated data relevant to ongoing land supply and capacity requires public-private cooperative arrangements.

Local Government Data Management: Administration Versus Policy Support

Much of the data collected and many of the databases developed by and for local governments have characteristics that do not serve land monitoring well. This is because "[these] information systems . . . are used for administration and [only] gradually are oriented toward policy analysis" (Bollens and Godschalk 1987, 321). For example, data on building and occupancy permits generally do not register whether actual construction or occupancy took place as permitted. Hence, a permit tracking system used to assess the outcomes of land use policies and plans as implemented should include data on both the permitting process and the outcomes of actions initiated through that process.

To reconcile the primarily administrative focus of public-sector functions with the needs of monitoring and planning analysis, local governments may have to position LSCM as a primary GIS user, perhaps even giving it a leading role in the development of shared GIS. However, LSCM system design will have to consider the different ways in which different departments collect and maintain their data—especially preferred spatial units and approaches to data updating—and adjust data management and analysis methods accordingly.

Engaging Multiple Jurisdictions

For reasons discussed elsewhere in this book—including the metropolitan scope of both land markets and growth management and monitoring mandates—land supply and capacity are most prominently regional concerns. Thus, land monitoring requires local governments to go beyond multidepartmental GIS data sharing to multijurisdictional coordination. Strong internal coordination within each jurisdiction using a GIS is a likely prerequisite to the success of regional coordination efforts.

The matrix in Figure 7.1 illustrates the multiplicity of data sharing relation-

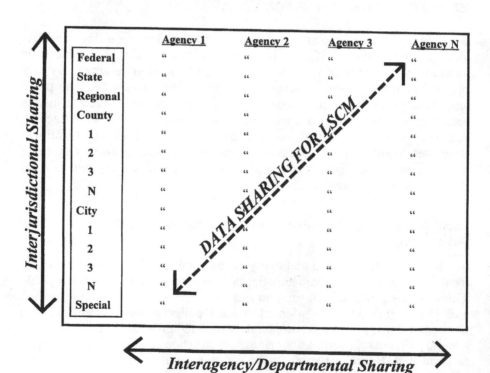

FIGURE 7.1 Data sharing relationships

ships that take place across a region's jurisdictions and departments. Notably, data sharing necessary for LSCM cuts across both dimensions. However, no guide exists for negotiating multijurisdictional information systems. This new territory for most local governments is further complicated by ever-changing technology (as in the development of web-based and networked systems) and by the dynamic division of labor and expertise between GIS "builders," "managers," and "users." A key to success is likely to be the constitution of a coordinating body representing all the institutional and related professional perspectives depicted in Figure 7.1.

Private Sector and Market Data

It is generally agreed that the private sector will continue its limited use of GIS and not actively seek to develop and maintain comprehensive land information systems. At the same time, increased pressure to include a private-sector perspective in LSCM, including a responsive treatment of land markets, suggests the desirability of incorporating private-sector data.

The difficulty of achieving this task first hinges on overcoming a track record of mutual distrust, and at times outright conflict, between the private and public sectors, particularly in the arena of growth management (see "Commentary" to Chapter 3 by Rolfe). Second is the need to resolve proprietary interests in data that are used privately to gain competitive advantage, or where there is a perceived need to protect significant past investments in database development. Technical issues loom as well—such as establishing standardized unique identifiers to link private-sector property information with public-sector parcel records. Finally, market data commonly collected for zip code areas have to be made compatible with public-sector data collected for other geographic units (e.g., census, analysis zones).

Efforts at coordination between public- and private-sector players in the arena of LSCM have generally not been successful, because they have lacked direction, leadership, and financial support. Dueker and DeLacey (1990, 490) suggest that planners are "perhaps in the best position to coordinate . . . [the] interaction among the participants in the land development process." Clear mandates to include market-related data and associated financial support will help to ensure that planners can fulfill this promise in the future.

OUTLOOK FOR THE FUTURE

The promises of greater data sharing and coordination in the realm of local government GIS are many, and, if realized, they will have significant impacts on the practice of LSCM. Indeed, the implementation of truly multipurpose, multiparticipant land information systems, especially those coordinated at a regional level, would resolve many of the data collection and sharing problems

associated with land monitoring. This chapter has concentrated on the requirements of enterprise-wide/multijurisdictional GIS, considering the data-sharing
issues particular to LSCM.

At this point in time, such highly coordinated GIS remains an ideal somewhat removed from reality. Significant questions persist regarding how to get
there and what the implications for future approaches to LSCM are.

Strategic Approaches to Data Sharing

As a first consideration, jurisdictions will have to plan carefully and strate-
gically the integration of their databases. Local governments with long-
established GIS programs may face considerable institutional inertia to change
in data management and sharing practices. According to Tosta (1998), "it is
primarily the larger jurisdictions, with a longer history of GIS, with larger
staffs, and in many cases, with more than one GIS installation, which are the
most challenged." For these cases, episodes of systemwide technology update
or overhaul are likely to provide occasions to redress effectively the persistent
fragmentation. The specific coordination of GIS for LSCM may have to wait
for such opportunities to become fully integrated within the normal information management structure of local government.

Jurisdictions engaged in redesigning their GISs or implementing new systems must, early in the process of system design, consider the needs of land
supply monitoring—including data quality, format, update procedures, and
accessibility. Owing to the complexity and expense of the endeavor, the reasons
for incorporating LSCM requirements must be clearly understood. Needed
are precisely stated objectives for the monitoring program to ensure that the
appropriate data are collected and the appropriate data-sharing structures
are installed. To this end, a formalized LSCM data needs analysis may be
undertaken, similar to the needs analyses that inform broader GIS design and
implementation efforts. In all cases, as well, consistent funding must be secured
early on for land supply database development and GIS coordination.

Given the fragmentation of GIS database management in local government,
LSCM may indeed serve as the linchpin for greater cooperation between
departments and jurisdictions. This possibility becomes not only real, but
also perhaps inevitable in light of increasingly prevalent mandates addressing
"consistency" (between public policies, regulations, and investments related
to land use) and "concurrency" (between private development and infrastructure provision).

As a second consideration, successful data sharing for or with LSCM involves rethinking centralized, top-down approaches to designing land monitoring systems, particularly considering GIS realities. There is a tendency among
even well-informed policymakers to underestimate database development
costs and to overestimate the capabilities of GIS in real-world settings. A lack
of technical knowledge at the higher levels of decision making often translates
into inadequate budgetary support for GIS database development. Conversely,

LSCM approached as a purely technical exercise may not adequately serve policy requirements in their full complexity. As collaborative efforts are carried out to implement LSCM, it is crucial that there be adequate representation by both technical and nontechnical staff and departments.

Decentralized Approaches to LSCM

A top-down approach to coordinating LSCM often conflicts with the realities of locally based planning. Regional planning processes often function from the bottom up, with each local jurisdiction following substantively different plans with different definitions of land use and different development objectives. Embracing the decentralized characteristics of most planning practice and GIS implementation suggests two ways in which approaches to LSCM could be adjusted. First, the confluence of specific interests and data requirements among the range of players or departments involved in local government and LSCM (outlined earlier and illustrated by the axes of Figure 7.1) presents an opportunity to establish formalized reciprocal data management and access agreements. Such agreements could structure LSCM data sharing by distributing responsibilities between the various departmental entities. Distributed responsibilities could, in turn, take two forms. On one hand, they could entail the provision of raw data that are mutually compatible with other land monitoring data. The data must be at an appropriate level of disaggregation and appropriate in content to the task of analyzing land supply and capacity. On the other hand, distributed responsibilities could entail the actual performance of components of the LSCM process by local agencies, providing subsequent inputs to comprehensive regional analyses. With some measure of standardization, for example, buildable land supply figures generated by local analyses could merely be aggregated to subregions or to the entire urbanized area. Such a distributed analysis model could take the following paths: public utilities or environmental agencies conducting preliminary work to delineate development constraints; assessors or economic development departments performing studies of land availability and market activity; or housing departments, along with private-sector interests, refining land use or housing data for use in fine-grained analyses of housing provision and affordability.

Finally, both local and regional policy may opt to carry out LSCM in a two-tiered manner. In such an arrangement, a regional body, such as a county or a metropolitan planning organization (MPO), would conduct some monitoring tasks separately from local efforts, with considerable reliance on locally generated data but likely depending on remote sensing and other large-area data collection techniques. Such regional monitoring would likely utilize coarse geographic units of GIS data and address long-term supply-and-demand balance (20 to 50 years). This would leave to local LSCM programs the role of carrying out short-to-medium-term supply tracking and analysis in greater detail. These local efforts would thus respond appropriately to private-sector concerns about adequate amounts of land for development, providing a broad range of feedback to local land regulators, and functioning as a check against systematic biases or inaccuracies in the regional estimates.

Whatever overall organizational settings are devised for LSCM in the future, they will be related intimately to local government GIS structures. The question is not which comes first, but how they are integrated. In this context, the next decade of technological change will continue to affect local government options for orchestrating the complexities of LSCM.

REFERENCES

Birkin, Mark, Graham Clarke, Martin Clarke, and Alan Wilson. 1996. *Intelligent GIS: Location decisions and strategic planning.* Cambridge, U.K.: GeoInformation International; New York: John Wiley & Sons.

Bollens, Scott A., and David R. Godschalk. 1987. Tracking land supply for growth management. *Journal of the American Planning Association* 3: 315–327.

Castle, Gilbert H. 1997. The bigger picture. *Business Geographics* 2:16.

———, ed. 1993. *Profiting from a geographic information system.* Fort Collins, Colo.: GIS World Books.

Charlton, Martin, and Simon Ellis. 1991. GIS in planning. *Planning Outlook* 1: 20–26.

Dueker, Kenneth J., and P. Barton DeLacey. 1990. GIS in the land development planning process: Balancing the needs of land use planners and real estate developers. *Journal of the American Planning Association* 4: 483–491.

Federal Geographic Data Committee, Cadastral Subcommittee. 1996. *Cadastral data content standard for the national spatial data infrastructure.* ⟨http://www.fgdc.gov/ standards/documents/standards/cadastral/cadstandard.pdf⟩ (October 10, 1997).

Fernandez-Falcon, Eduardo, James R. Strittholt, Abdulaziz I. Alobaida, Robert W. Schmidley, John D. Bossler, and J. Raul Ramirez. 1993. A review of digital geographic information standards for the state/local user. *URISA Journal* 2: 21–27.

Godschalk, David R., Scott A. Bollens, John S. Hekman, and Mike E. Miles. 1986. *Land supply monitoring: A guide for improving public and private urban development decisions.* Boston: Oelgeschlager, Gunn & Hain, in association with the Lincoln Institute of Land Policy.

Heikkila, Eric J. 1998. GIS is dead: Long live GIS! *Journal of the American Planning Association* 3: 350–360.

Kaiser, Edward J., David R. Godschalk, and F. Stuart Chapin. 1995. *Urban land use planning.* Urbana and Chicago: University of Illinois Press.

Kollin, Cheryl, Lisa Warnecke, Winifred Lyday, and Jeff Beattle. 1998. Growth surge: Nationwide survey reveals GIS soaring in local governments. *GeoInfo Systems* 2: 25–30.

Nedovic-Budic, Zorica, and Jeffrey K. Pinto. 1998. *Coordinating development and use of geographic databases.* Unpublished project report. Champaign: University of Illinois.

Public Technology, Inc., Urban Consortium, and International City Management Association. 1991. *Local government guide to geographic information systems: Planning and implementation.* Washington, D.C.: Public Technology, Inc., Urban Consortium, and International City Management Association.

Thrall, Grant. 1997. Telephone interview by author. November 19.

Tosta, Nancy. 1998. Presentation at May 1998 Parcel-Base GIS for Land Supply and Capacity Monitoring Conference, University of Washington, Seattle.

von Meyer, Nancy, and D. David Moyer. 1994. New wheels for a new highway: Land information standards. *URISA News* 141: 1, 6

COMMENTARY: HIGHLIGHTS OF INSTITUTIONAL AND ORGANIZATIONAL ISSUES

Zorica Nedovic-Budic

Advances in technology seem to evolve faster than the institutional and organizational capacity to adjust to them and absorb them. Although technological solutions carry with them the potential to revolutionize the planning, management, and administration of urban development, organizational capacities and structures remain unprepared for the new forms and processes required for effective utilization of planning and decision support systems (Godschalk et al. 1986; Bollens and Godschalk 1987). This situation slows progress in managing and guiding urban growth and redevelopment.

Chapter 7 is on target with its assessment of the state of the art. From the late 1980s to the present, GIS technology has been widely diffused in local government settings, and among planning agencies in particular (Budic 1993; Warnecke et al. 1998). One of the main reasons for planning agencies to seek adoption of GIS technology is its capability to integrate graphical and tabular data from various sources and across different areas. Many GIS functions match the day-to-day needs of planning practice. Unfortunately, in the face of those needs, there is little coordination among organizations and jurisdictions, and GIS and database developments remain fragmented even within single organizations (Craig 1995; Tosta 1995). Institutional inertia and concern about privacy of data drive this lack of coordination among public agencies and make the inclusion of the private-sector data difficult.

Barriers to coordination also stem from GIS design and implementation practices, which often treat *operational, management,* and *policy* information as separate factors. Huxhold (1991) recommends integrating these three levels by building GIS utility from the bottom up, through continuous data collection at the operational level. This is a strategy that, in the context of land monitoring, can be implemented by linking "transactional" updating to administrative land management "actions."

As suggested by Moudon and Hubner, three core tasks emerge to ensure that local governments fully exploit advances in technology for the purposes of LSCM: (1) identifying and motivating the agencies and jurisdictions to be included in interorganizational activities aimed at coordination and cooperation in developing GIS and sharing spatial data needed for LSCM, (2) finding the most effective techniques and strategies for achieving successful interorganizational activities, and (3) understanding the changes required for incorporating the new technology into organizational processes and for improving the flow of information between organizations. These tasks, in turn, require understanding (1) the *scope* of the multiparticipant system, (2) the internal organizational and interorganizational *contexts*, (3) the *interaction* necessary for undertaking coordinated GIS and database developments, and (4) the *organizational changes* required to mutually adjust the new technology and organizational processes. These latter four components are discussed in the following sections.

Scope

The multiparticipant and, possibly, multijurisdictional setting described as necessary for comprehensive LSCM is most complex and therefore most difficult to accommodate. It requires integration of data, first, within each jurisdiction and, second, across jurisdictional boundaries. This most complex situation can be considered as one extreme in a range of solutions and configurations that can support an LSCM program. Depending on the scope of the program, the supporting information can be provided by as few as two agencies or by a large number of agencies. In considering the scope of the LSCM program as a series of steps, four levels of participation can be differentiated, each level referring to a broader scope of LSCM and encompassing a larger group of organizations:

- *Planning and land records/tax assessment.* Nedovic-Budic and Pinto (1998) found that attracting the main player is the key for successful interorganizational partnerships. The land records management and tax assessment offices are the primary participants that must be involved in building a PBGIS and in data sharing. Hence, relationships must be established, first between planning departments and land records and tax offices within each jurisdiction, and then across jurisdictions.
- *Infrastructure providers.* The next step is to coordinate and share databases with public and private agencies in charge of infrastructure—mainly roads, sewer, and water provision.
- *Private sector.* The third step is to interact with private-sector parties concerned with land development and transactions (i.e., real estate agencies and developers). Their activities are independent from, yet closely related to and affected by, local planning and regulatory activities.
- *Other agencies.* Finally, community development, housing, economic, and environmental information can be obtained from a variety of primary

and secondary sources and is often available within planning agencies. This information contributes to a comprehensive view of urban development activities and their impact.

For effective LSCM, the integration of GIS and databases within each organization and jurisdiction, and across several organizations and jurisdictions participating in the program, may range in intensity from the development of coordinated databases to the joint acquisition and use of hardware, software, personnel, and space or facilities. Development of joint applications and functional integration are the two most complex forms of integration. Whereas coordinated and shared databases are a prerequisite for local and regional LSCM activities, it is the *functional integration* of organizational processes that will result in effective monitoring and guidance of urban development. Therefore, it is useful to relate the LSCM program requirements closely to the mission and functions of each relevant organization. Specifically, the challenge lies in the likely mismatch, which the chapter highlights, between the urban planning, management, and administrative functions, in which the first government function can happen at the regional level but the latter two are most often performed at the local (internal or single jurisdiction) level.

Context

Understanding the context is crucial for establishing interorganizational GIS and database sharing (Nedovic-Budic and Pinto 1999). Azad's (1998) case studies show that "working the context" is necessary for success in enterprise-wide GIS implementation. The contextual issues qualified as the most significant determinants of success are (1) organizational interdependence and (2) motivation. The nature of interdependence determines the structurability of the relationships, as well as the range of coordination mechanisms that can be applied to ensure a functional relationship. Organizations relate to each other in ways that reflect *pooled, sequential,* or *reciprocal interdependence* (Thompson 1967). Kumar and van Dissel (1996) suggest that pooled interdependence requires coordination by standardization; for sequential interdependence, it is appropriate to apply coordination by plan; and with reciprocal interdependence, coordination is pursued by mutual adjustment. The structurability of the relationship decreases from high to low in moving from a simple pooled relationship toward reciprocal interdependence. Comprehensive land supply and capacity monitoring most likely implies a complex interdependence that is not easy to structure.

Meredith (1995) postulates that when organizational interdependence already exists, there is lower resistance to interorganizational activities. Focusing on the minimal partnership necessary for establishing parcel-based GIS, we proceed to examine the interdependence between planning agencies and land records management and assessment offices. So far, the relationships between these agencies have been mostly one-way: Land records and tax assessment offices generate data that are essential for planning operations; thus, planning agencies are dependent on these offices. Some reciprocity is also evident: Land records and tax-related databases often include information derived from planning implementation documents, such as zoning ordinances and maps. Although Moudon and Hubner argue that the land records and tax assessment offices would benefit from more detailed and precise information on land supply and capacity, their dependence on planning data should be further clarified. This leads to the question of motivation: Reasons must be found for land records and tax assessment agencies to provide information in the format and with the content, accuracy, and timeliness needed by planning agencies above and beyond that regularly available through access to government records.

Generally, land records management and tax assessment offices are quite independent in their operations, making their voluntary support for LSCM databases and programs unlikely. Given this situation, several alternative approaches may be entertained to secure more cooperative participation: (1) mandates from a governmental or intergovernmental body (Cummings 1980), (2) a sense of duty to respond to informal requests or interest expressed by local authorities (e.g., an elected official or higher-level administrator), (3) common interests among organizations that value the same goals, or (4) exchange inducements when returns are expected or received (O'Toole and Montjoy 1984). Land monitoring agencies may pursue these alternatives, for example, by convincing a county commissioner that interorganizational cooperation and GIS sharing is crucial for the well-being of the region, and by having the commissioner promote those activities either through an official mandate or through political pressure. Agencies may also work with land records management and tax assessment officials to arrive at a common goal—for instance, raising local property tax revenues. Finally, and probably most effectively, they may offer resources (e.g., staff, financial) in support of special contributions and responsiveness to LSCM program needs by other agencies.

The motivation-related consideration applies across the organizations listed as relevant in the LSCM activities. A general rule for understanding this motivation is that the more independently organizations function in their own operations, the less interested they are in coordination or cooperation with other organizations and jurisdictions.

Interaction

To coordinate their GIS and database activities, organizations interact with each other at various levels and for a variety of purposes. Frequent and open *communication,* persistent negotiation and bargaining, commitment, and teamwork are all qualities of successful interaction. Moudon and Hubner point out correctly that guides for interaction are lacking. Guidelines are, however, important for helping multiple organizations achieve mutual PBGIS-

and LSCM-related goals. Negotiations are often necessary because the participant agencies are likely to operate from different standpoints and have unique interests and needs. In addition, the level of familiarity with the use of GIS technology usually varies among organizations and among their units, and the GIS adoption landscape across metropolitan regions is quite diverse.

Communication involving heterogeneous groups is a challenge that requires the creation of a common working language. It is often necessary to identify semantic differences and commonalties between concepts held by participants, particularly when they may have multiple interpretations of technical questions, terms, or data in use (Harvey 1997). LSCM is certainly an arena in which those differences and commonalties can be explored and resolved. Terms such as *land use, land supply and capacity, open space conservation, urban boundaries, built-up areas,* and many others, have different meanings for planners, tax assessors, public workers, and environmentalists. Clarifying and understanding diverse terminology is the first step toward successful joint GIS activities. Agreeing on common meanings is the next step to facilitating interorganizational communication and coordination.

Coordinating GIS and database activities takes extra time and energy and constitutes a possible resource drain (Brown et al. 1998). Nedovic-Budic and Pinto's (1998) cases show that transaction costs are associated with maintaining relationships between member organizations that are differentially committed to GIS and database-sharing arrangements. Transaction costs are the costs of managing interactions while minimizing opportunistic behavior (Kumar and van Dissel 1996). Critics of multiparty implementation efforts claim that partnerships are likely to experience circumscribed coordination, limited scale economies, frequent delays, and problematic outcomes. These problematic outcomes are particularly likely to occur when the scope and goals of a project are not clearly specified and when interactions are "polluted" by private agendas, interorganizational conflicts, and unrelated coalition building. On the other hand, shared understanding, trust, mutual credibility, and willingness to compromise all lead to successful interactions (Citera et al. 1995).

Organizational Change

The implementation of a parcel-based GIS for effective LSCM will ultimately require more than shared spatial databases, hardware, software, or personnel. It will demand organizational change. Organizational change involves restructuring to create new organizational forms, processes, procedures, information flows, and responsibilities, which correspond to and are enabled by new technology. The purpose of such change is to integrate technology and organization by mutual adjustment and to institutionalize the activities needed to use the technology effectively in pursuing organizational missions and functions (Nedovic-Budic 1997).

Alexander and Randolph (1985) found that the fit between organization and technology is a better predictor of organizational performance than technology, structure, or technology and structure combined. These authors developed a simple measure of the organization-technology fit, which suggests matching mechanistic structures with routine technologies and organic (flexible) structures with nonroutine applications. Gorry and Morton (1989) view the fit to depend on the type of decisions (structured, semistructured, and unstructured) and their respective use for operation, control, and strategic planning. The challenge in fitting parcel-based GIS with a LSCM program across several organizations has to do with the nature of the tasks involved, which include both structured and unstructured decisions and require routine and nonroutine applications. For example, retrieving and summarizing information on development permitting activities and the amount of land affected within a certain time period (e.g., daily or weekly) is a routine application that can be handled with standardized database procedures, query interfaces, and reports. Evaluation of land development strategies is, on the other hand, based not only on the facts about the impact on land supply and building density, but also on the attitudinal information about quality of life implied by various built environments, the importance of protecting agricultural land and conserving natural resources, the tolerance for environmental pollution and traffic congestion, and other issues that cannot be as easily automated within an information or decision-support system.

The kind of organizational change necessary to support LSCM will depend on the scope and the purposes of monitoring—organizational change will likely be simpler in the case of basic land inventories than in comprehensive supply and capacity assessment involving concurrent consideration of infrastructure, services, or social and environmental impacts. At a minimum, planning agencies will have to establish links with land records management and tax assessment offices, zoning administration, and building permitting units. These "links" involve a true (i.e., not superficial) integration of organizational processes, functions, personnel, and resources. The changes associated with such links will likely generate impacts across the organization—indeed, the uncertainty related to the anticipated impact may be sufficient to stall the integration efforts. As discussed earlier, the degree of integration will depend on the motivation of the units involved. A horizontally and vertically distributed and integrated LSCM system—one that crosses jurisdictional boundaries and involves a variety of organizations, such as utility companies, real estate agencies, environmental protection groups, and open space management organizations—will be even more challenging to achieve.

In the private sector, "business process change" has become a standard procedure accompanying the introduction of new technologies. Methodologies, techniques, and tools for facilitating change of this kind are intensively studied in the field of management information systems (Kettinger et al. 1997). In contrast, organizations in the public sector are slow to adapt to new technologies and to undergo change. Organizational inertia, therefore, is a major cause of the mismatch between organizations and technology and, consequently, for the failures in introducing GIS (Craig 1995; Tosta 1995).

Nedovic-Budic and Pinto (1998) report very little organizational change detected in five cases studied to explore coordination of geographic information systems and databases among local and regional organizations. As noted, poor institutional capacity for adjustment and change has been a persisting problem in applying automated land supply and capacity monitoring systems since the early 1980s.

An evolutionary approach to organizational change may appropriately anticipate an increase in sophistication among land information systems as they move from mapping function to decision-making support. GIS applied to land management and planning will improve with continuous use over a long period of time. It is, therefore, too early in the general evolution of GIS in public agencies to draw any firm conclusions about its performance. Better understanding of GIS implementation factors and processes within various organizational settings and configurations will help local governments devise effective strategies to achieve more quickly the desired impacts from GIS technology. General lessons from the application of information technologies are available and useful, but, unfortunately, they are rarely consulted by organizations engaged in GIS implementation (Budic and Godschalk 1994). Eason (1988), for instance, provides a detailed description of sociotechnical systems design and describes five implementation strategies that relate directly to organizational contexts; these range from revolutionary instantaneous change to phased and incremental evolutionary approaches.

There are also examples of advanced GIS-based integrated systems that automate the entire planning and building process and link several applications within a GIS environment. In Alberta, Public One-Stop Service (POSSE), developed by the City of Edmonton Planning and Development Department (Caldwell et al. 1998), employs a "parallel running" strategy utilizing temporary manual and computerized systems that operate simultaneously. It involved integration and enhancement of several previously existing systems (including socioeconomic land use tracking, financial and document management, tax collection, and land titles) and pursued new business processes among the relevant branches of the municipal government. POSSE development was based on extensive analyses, but retained flexibility and openness to new demands for change. Strong project leadership, a committee structure, teamwork, and public/private partnerships are the pillars of POSSE's success.

Another example, the Mississauga (Ontario) Approvals Xpress (MAX) System was developed by the City of Mississauga's Information Technology Division (Alley 1997). This system integrated four separate legacy systems used to track building permits, planning applications, properties, and developable land. MAX automated more than 200 activities, related to the processing of applications, within a seamless system. Careful design of user interfaces and links with other systems allowed MAX to diffuse extensively throughout Mississauga's municipal government agencies. The system designers considered the determination of property definition one of the most complex issues they had to resolve in working with all participants in the system development.

Both POSSE and MAX received the Urban and Regional Information System Association's (URISA) Exemplary Systems in Government Award. Because these systems are local and interdepartmental in nature, they do not satisfy the regional and multijurisdictional setting suggested for LSCM. They do, however, provide a step ahead in tackling the most challenging institutional and organizational issues that pervade multijurisdictional efforts as well. Both systems attempted and achieved major organizational changes that transformed the way land development management business was done.

References

Alexander, J. W., and W. A. Randolph. 1985. The fit between technology and structure as a predictor of performance in nursing subunits. *Academy of Management Journal* 4: 844–859.

Alley, J. 1997. The City of Mississauga's Mississauga Approvals Xpress (MAX) System. URISA Exemplary Systems in Government Application. Mississauga, Ont.: City of Mississsauga, Information Technology Division.

Azad, B. 1998. Management of enterprise-wide GIS implementation: Lessons from exploration of five case studies. Ph.D. diss., Massachusetts Institute of Technology.

Bollens, Scott A., and David R. Godschalk. 1987. Tracking land supply for growth management. *Journal of the American Planning Association* 3: 315–327.

Brown, M. M., L. J. O'Toole Jr., and J. L. Brudney. 1998. Implementing information technology in government: An empirical assessment of the role of local partnerships. *Journal of Public Administration Research and Theory* 8: 499–525.

Budic, Zorica D. 1993. GIS use among Southeastern local governments—1990/1991 mail survey results. *URISA Journal* 1: 4–17.

Budic, Zorica D., and David R. Godschalk. 1994. Implementation and management effectiveness in adoption of GIS technology in local governments. *Computers, Environment and Urban Systems* 5: 285–304.

Caldwell, R., J. Mines, and D. Fraser. 1998. POSSE—Moving from the paper world to the electronic world. *Plan Canada* 9: 12–17.

Citera, M., M. D. McNeese, C. E. Brown, J. A. Selvaraj, B. S. Zaff, and R. D. Whitaker. 1995. Fitting information systems to collaborating design teams. *Journal of the American Society for Information Science* 7: 551–559.

Craig, W. J. 1995. Why we can't share data: Institutional inertia. In *Sharing Geographic Information.* Harlan J. Onsrud and Gerard Rushton, eds. New Brunswick, N.J.: Center for Urban Policy Research, 107–118.

Cummings, T. G. 1980. Interorganization theory and organizational development. In *Systems theory for organization development.* T. G. Cummings, ed. New York: John Wiley & Sons.

Eason, K. D. 1988. *Information technology and organisational change.* London: Taylor & Francis.

Godschalk, David R., Scott A. Bollens, John S. Hekman, and Mike E. Miles. 1986. *Land supply monitoring: A guide for improving public and private urban development decisions.* Boston: Oelgeschlager, Gunn & Hain, in association with the Lincoln Institute of Land Policy.

Gorry, G. A., and M. S. S. Morton. 1989. A framework for management information systems. *Sloan Management Review* (spring): 49–60.

Harvey, F. 1997. Improving multi-purpose GIS design: Participative design. In *Spatial information theory*. S. C. Hirtle and A. U. Frank, eds. Berlin: Springer Verlag.

Huxhold, William E. 1991. *An introduction to urban geographic information systems*. New York: Oxford University Press.

Kettinger, W. J., J. T. C. Teng, and S. Guha. 1997. Business process change: A study of methodologies, techniques, and tools. *MIS Quarterly* 1: 55–80.

Kumar, K., and H. G. van Dissel. 1996. Sustainable collaboration: Managing conflict and cooperation in interorganizational systems. *MIS Quarterly* 3: 279–300.

Meredith, Paul H. 1995. Distributed GIS: If its time is now, why is it resisted? In *Sharing Geographic Information*. Harlan J. Onsrud and Gerard Rushton, eds. New Brunswick, N.J.: Center for Urban Policy Research, Rutgers, The State University of New Jersey, 7–21.

Nedovic-Budic, Zorica. 1997. GIS technology and organisational context: Interaction and adaptation. In *Geographic information research—Bridging the Atlantic*. Massimo Craglia and Helen Couclelis, eds. London: Taylor & Francis, 165–184.

Nedovic-Budic, Zorica, and Jeffrey K. Pinto. 1998. *Coordinating development and use of geographic databases*. Unpublished project report. Champaign: University of Illinois.

———. 1999. Understanding interorganizational GIS: A conceptual framework. *URISA Journal* 1: 53–64.

O'Toole, L. J., Jr., and R. S. Montjoy. 1984. Interorganizational policy implementation: A theoretical perspective. *Public Administration Review*. (Nov./Dec.): 491–505.

Thompson, James D. 1967. *Organizations in action*. New York: McGraw-Hill.

Tosta, Nancy. 1995. The evolution of geographic information systems and spatial data-sharing activities in California state government. In *Sharing geographic information*. Harlan J. Onsrud and Gerard Rushton, eds. New Brunswick, N.J.: Center for Urban Policy Research, 193–206.

Warnecke, L., J. Beattie, K. Cheryl, W. Lyday, and S. French. 1998. *Geographic information technology in cities and counties: A nationwide assessment*. Washington, D.C.: American Forests.

8

Simulating Land Capacity at the Parcel Level

Paul Waddell

Land capacity analysis focuses mostly on the attributes of the land and its potential uses. This orientation examines the supply side in detail, whereas analysis of demand remains on such an aggregate level that it serves only to determine a probable time horizon during which the existing capacity will be sufficient for urban development. Hence, although land capacity analysis examines vacant or underutilized parcels, the land use plan, zoning, and the presence of infrastructure, it has done little more than acknowledge the existence or relevance of consumer preferences (see Mildner et al. [1996] and Chapter 2). Herein lies its greatest challenge.

The composition of urban land consumers, and their preferences, are critical to land capacity analysis, since the likelihood of a parcel of land being developed for a particular use and at a particular density is a function of consumer preferences. Land capacity is as much a function of demand for land as it is of land supply. Regulatory frameworks may restrict development options for a parcel to a subset of land uses, and may impose minimum or maximum density limits, environmental constraints, and other land regulations that further restrict development options. Ultimately, however, the specific use of the

Monitoring Land Supply with Geographic Information Systems, edited by Anne Vernez Moudon and Michael Hubner ISBN 0 471371673 © 2000 John Wiley & Sons, Inc.

parcel is driven by the profitability of alternative development outcomes on the parcel, which in turn derives from the tradeoff between development costs and the consumer's willingness to pay for the development.

Three aspects of consumer preferences are salient to this discussion: the composition, or mix, of consumers; consumer preferences for various types of development; and consumer preferences for different locations, or submarkets. At the aggregate level, if the composition of consumers changes (e.g., through relatively large inmigration of affluent or poor households), then the demand for certain types of development will grow disproportionately, such as the current growth in demand for large-lot single-family housing or for condominiums. As for commercial development, different types of businesses have differing needs for development as well, with many factors influencing these choices. Moreover, preferences for different development types and densities may be changing over time. Finally, some of the development options being considered in land use plans call for forms of development in which consumer preferences are not well understood, such as dense urban villages and neotraditional neighborhood design.

At the submarket level, consumer preferences translate into disproportionate demand for housing or nonresidential space in particular geographic areas, on the basis of accessibility to desired amenities and activities. Failure to recognize this locational bias in preferences is perhaps one of the most serious oversights in current land capacity analysis. It explains why certain areas that have substantial land for development remain underdeveloped while others that have insufficient land continue to develop, even with rapidly increasing prices. An example of this problem is industrial land supply, which, if zoned in the "wrong" places within a metropolitan area, will render the aggregate supply of such space meaningless in terms of its long-term adequacy to accommodate development.

URBAN SIMULATION MODELING

Recent innovation in the development of operational urban simulation models provides a potential avenue for addressing some of the concerns about land capacity analysis raised in the preceding section and in earlier chapters. The key omission in most land capacity analysis is a detailed analysis of variations in consumer demand for housing by type and location. Urban simulation models such as UrbanSim provide a mechanism to model the demand for alternative types of development at detailed spatial locations, so that the market effects on the consumption of available land can be more effectively considered in refining estimates of land capacity under alternative land policies. The potential application of urban modeling to land capacity analysis is probed in this section by examining the development and application of the UrbanSim model in Eugene-Springfield, Oregon.

The UrbanSim model was developed as a prototype metropolitan land use model for the Oregon Department of Transportation and intended for use by metropolitan planning organizations (MPOs) in the state, in conjunction with the MPO travel demand forecasting models (Waddell, 1999; 1998a, b). The motivation for the model development project was to facilitate land use and transportation planning at the regional level within the context of growth management policies at the state and local levels. The initial prototype has been completed, and the design of the second-generation model is under way. The model is being implemented in Honolulu, Salt Lake City, and Seattle.[1]

The model can be summarized as a behavioral simulation of the choices made by key actors in the urban development process: households, businesses, developers, and governments. It operates in a quasi-dynamic manner, over annual increments in time. The following sections present an abbreviated description of the components of the model relevant to this discussion: consumer demand, the land development process, the role of land regulations and transportation infrastructure, and the use of parcel-level data.

The Demand Side

The demand side of UrbanSim simulates consumer preferences of households and businesses by market segment for locations and development types. Households are stratified by income, age of the head of the household, number of persons, and presence of children. Businesses are stratified by industry and number of employees. For different market segments, we estimate consumer preference for development types and locations by observing the choices made by recent movers and using these consumer choices to estimate statistically the preferences of a group of consumers.

The prices paid by consumers for the development types at the chosen locations represent consumers' "bids" for these developments. We know that the observed locations of recent movers are the results of their successful bids for the properties, meaning that they were the highest bidders for the particular developments they now occupy, based on the assumption that landowners sell to the highest bidder. The examination of successful bids of a sample of recent movers, including the prices paid and the characteristics of the development types and locations, forms the foundation for the demand analysis. We label these consumer preference estimates "bid functions." The characteristics considered in the bid functions are those that we anticipate from theory to influence consumption choices, and that we can measure using available data.

The variables considered in the household bid function in the Eugene-Springfield application of UrbanSim include the following:

[1] Funding for UrbanSim development has come from the Oregon Department of Transportation, the Oahu Metropolitan Planning Agency, the Governor's Office of the State of Utah, the National Cooperative Highway Research Program 8-32(3), Integration of Land Use Planning and Multimodal Transportation Planning, and National Science Foundation grant CMS-9818378.

- Housing types (using single-family residential as the omitted comparison type)
- Access to jobs and shopping from the zone
- Density, age, and size of the housing stock of each type in the zone
- Income mix of residents in the zone and the percentage with children
- Land use composition
- Travel time to the central business district

Note that each of these variables must be updated as part of the simulation; they are endogenously predicted within the model. The variables represent characteristics of housing and location anticipated from urban economics and geography to influence residential location. The bids are for groupings of individual housing units by traffic analysis zone and housing type.

The estimation of the bid function for households of different types yielded results in the Eugene-Springfield application that were reasonable and intuitive as well as statistically significant. The residential bid functions were estimated separately for households stratified by income group and presence of children, generating eight different models. The results of these estimations, discussed in detail in Waddell (1998b), are summarized here.

The model results explain between roughly 70 and 80 percent of the variation in observed housing prices, indicating that the approach is reasonably robust. This explanatory power is relatively high, considering that the bid estimates are derived from assessed values rather than sales transactions, and that the household data were created from a mixture of census, parcel, and transportation data (Waddell 1998a). These results are indicative of the specification of consumer preference functions that could be useful in relation to assessing land capacity as well.

The results also reveal preferences that differ between household types in interesting ways, such as the relatively lower discounting for density by higher-income households without children, suggesting a potential market segment that would consider dense, in-town housing with substantial amenities. On the other hand, the most affluent households revealed a strong bias toward newer housing, as shown by the strong discounting for housing age. Multifamily and residential two- to four-unit housing types were significantly discounted by all consumer groups, as compared with single-family residential development.

This consumer preference framework is extended to predict the successful bidder for each location and development type by using the concept of consumer surplus, defined as the consumer bid minus the market price of an alternative. This theory holds that the probability that a particular consumer will be the highest bidder for a location is proportional to the bidder's consumer surplus. If we take the market price of an alternative as an indication of the highest bid among all consumers, then the likelihood that a particular consumer will be the highest bidder is proportional to the probability that his or her bid exceeds the bids of all other consumers. This approach solves the simultaneous problems of translating consumer preferences expressed in the bid function into observed location choices, and of assigning individual properties to the highest bidders.

The probability that a particular consumer will make a particular location choice is, then, a function not only of his or her preferences and budget constraints, but of those of all other consumers as well. This means that as the composition of consumers changes—for instance, toward higher-income households—we can expect the land market to change in predictable ways, with greater demand for the types of alternatives chosen by these consumers and higher prices as a result (holding supply and all other factors constant).

The specifications of bid functions for businesses were similar to those for households, but for the sake of brevity are not discussed here. A more complete description of the model and the results of the Eugene-Springfield application are available elsewhere (Waddell, 1998a; see also ⟨http://urbansim.org⟩).

The Supply Side

The supply side of the model attempts to simulate the development activity of private developers. The development model is based on a microsimulation of the expected profitability from the development or redevelopment of individual land parcels. Whereas the demand side groups parcels into clusters of the same development type within each zone, the development component treats each parcel individually. The reason for this approach is that the development process, whether for vacant parcels or through redevelopment of already developed parcels, is most understandable, and therefore the most straightforward to model, at the parcel level.

Within each parcel, we maintain in the model an accounting of the characteristics of land parcels, including the following:

- Acreage
- Land use
- Land value
- Improvement value
- Housing units
- Square footage of nonresidential development
- Age of improvements

These characteristics are updated by simulated new construction and redevelopment activity. In addition, the following characteristics of parcels may be designated by the user through geographic information system (GIS) processing and specific values assigned through the user interface to the model:

- Land use plan designation
- Environmental overlays:
    Wetlands
    Floodplains
    Slope
- Urban Growth Boundary
- Development costs

These latter characteristics form the bases for policies that restrict developer options for a particular parcel or alter the profit calculation by affecting development costs. Overlays such as wetlands, floodplains, and high slope areas are integrated with the parcel boundaries using polygon overlay operations in the GIS. These operations actually subdivide a parcel that is intersected by an overlay that crosses the parcel, such as a floodplain, creating separate subparcel records if a parcel is bisected by such a boundary. An alternative approach is to use the centroid of a parcel for the overlay and assign the entire parcel a value of the overlay (e.g., inside or outside the 100-year floodplain), but this approach would lose useful information for large parcels that contain some environmentally sensitive land.

The Urban Growth Boundary (UGB) is also overlaid on parcels, as are the metropolitan land use plan designations. These layers provide a basis for the user to interpret the land use plan and UGB policies directly and to apply rules that constrain developer behavior. User-generated rules include, for each land use plan designation, the types of development allowed and the minimum or maximum densities associated with each development type. In addition, development costs associated with policies, such as development impact fees to extend services, may be assigned to parcels or development types within the user interface.

Building types are classified from the land use classification codes available within the assessor parcel file. In the application of the model to Eugene-Springfield, the land uses were grouped as follows.[2]

- Residential single-family
- Residential two- to four-unit
- Residential multifamily
- Industrial
- Warehouse
- Retail

---

[2] These land use grouping will vary, depending on the location and the needs of the analysis. An effort is being made in the application of the model to Salt Lake City, for example, to develop a land use grouping that can be related to New Urbanism development types designed by Peter Calthorpe.

- Office
- Special purpose

Mixed-use categories existed only in the land use plan, but not in the Eugene-Springfield grouping of existing land uses. Given these data limitations, mixed land uses were accommodated in the model within a cluster of parcels rather than within a single parcel.

The actual behavior of the developer model is based on microsimulation of alternative development projects on individual parcels. The model checks the land use plan designation for the parcel, determines which development types are allowed and within what range of density, checks for environmental or UGB constraints, and then develops potential projects for each allowed option. The expected profitability of each of these tentative projects is estimated, and all of the projects across all parcels are rank-ordered by profitability.

The profit calculation uses the expected revenue from each potential project, based on the market value of the development type at the location in the previous year, and subtracts the costs of development. The development costs include the land cost, hard construction costs, soft costs of development (e.g., development impact fees), and, in redevelopment situations, the cost of existing improvements and the cost of demolishing those improvements.

$$\Pi_i(lb) = R_{lb}Q_{ib} - L_iA_i - H_bQ_{ib} - S_{lb}Q_{ib} - I_{ib}Q_{ib} - D_{ib}Q_{ib}$$

where:

Π_i (lb) is the expected profit from developing parcel i in location l into building type b

$R_{lb}Q_{ib}$ is the expected revenue from selling the project to consumers

L_iA_i is the land cost of parcel i (land cost per acre times acres)

H_bQ_{ib} is the "hard" construction cost of the development project (its replacement cost)

$S_{lb}Q_{ib}$ is the "soft" construction cost of developing the project, including development fees

I_{ib},Q_{ib}, is the cost of existing improvements on parcel i if it is being redeveloped

D_{ib},Q_{ib}, is the demolition cost for any improvements on parcel i if it is being redeveloped

The expected revenue is based on the current market price for space of building type b in location l:

$$R_{lb} = P_{lb}$$

The quantity of construction, Q_{ib}, is a function of the size of the parcel being evaluated and the density of construction. The density at which new construction occurs is predicted to be responsive to land prices, with higher land prices prompting capital/land substitution by developers. As land prices increase, we would expect developers to build at higher densities.[3] In markets with a supply of vacant land that is low relative to the demand generated by economic growth, vacant land prices should increase, sending an economic signal to developers to increase density. The density on which the expected profit of each development project is computed, then, is as follows:

$$\Phi_{lb} = \alpha_b + \beta_b \ln(P_{lb})$$

where:

Φ_{lb} is the density for parcels in location l and building type b

P_{lb} is the land price per acre in location l for building type b

α_b and β_b are estimated parameters

Market Price Adjustment

Land market price adjustment is an aspect of the model system that reconciles demands for land by households and business establishments with each other, and with the available supply of developed space in every year. It handles the assignment of moving businesses and households to the best (highest consumer surplus) alternative available and adjusts land prices according to the ratio of demand-to-supply in each zone. Because prices enter the location choice equations for businesses and households, an adjustment in prices will alter location preferences in the subsequent year, causing higher-priced alternatives to be more likely to be chosen by occupants who have higher incomes or less sensitivity to price, all else being equal. Similarly, any adjustment in land prices alters the preferences of developers in building new construction, in regard to type of space and density of construction.

Once households and businesses have evaluated all available alternatives and expressed their preferences (through a probability prediction in the location choice models), the simulation attempts to place households and businesses into buildings in proportion to their predicted probabilities. Alternatives that become "full" during this operation are removed from the rest of the allocation process, and households and businesses that are unable to locate within a building of highest utility to them are forced to accept the next-best alternative. This process is repeated until all businesses and households are assigned to locations.

---

[3] There is an important exception to this expectation: large-lot luxury single-family housing. Affluent households may bid a premium to retain low density, thereby reversing the general pattern. The degree to which this pattern exists can be determined from the data.

The market-clearing mechanism, then, is not strictly based on a full equilibrium price adjustment, in which perfect information exists and transaction costs are zero, so that prices for all buildings at each location adjust to the equilibrium solution that clears the market. Rather, the solution is based on an expectation of incomplete information and nontrivial transactions and search costs, so that movers obtain the most satisfactory location available, and prices respond at the end of the year to the balance of demand and supply at each location.

Once the market assignment is completed, the information generated by the market simulation about the relative demand and supply of each building type at each location is used to update prices. The magnitude of the price adjustment is based on current vacancy rates in the zone and region, as compared with a "normal" or "threshold" vacancy rate assumed to be at equilibrium. The adjustment factor is capped at an annual change of no more than a user-specified percentage in either direction.

The supply of housing and commercial space consumed in any iteration comes from existing vacant structures plus any new construction and redevelopment of structures that has occurred in the most recent period. New construction in a forecast interval can include committed, proposed, and potential development projects identified by the user as development events.

The form of the price adjustment is as follows:

$$P_{lbt} = P_{lbt-1} \frac{(1 + \alpha_b - V_{lbt}) + \lambda(1 + \alpha_b - V_{bt})^\beta}{1 + \lambda}$$

where

P_{lbt} is the land price of building type b in location l in year t

P_{lbt-1} is the previous year closing land price for the same building and location

V_{lbt} is the current vacancy rate for space in the building type and location l

α_b is the normal vacancy rate for building type b

β is a scaling parameter for the price adjustment, initially set to 1

λ is a parameter for weighting the regional and zonal influence

Because vacant land price is a key determinant of the profitability of alternative development outcomes on each vacant parcel (entering the cost side of the profit equation), the model must update vacant land prices as urban development proceeds, and there are changes in the prices of land developed around each vacant parcel. We expect that vacant land prices will adjust in relationship to developed land prices in a local area as a result of land speculation. Speculators purchase vacant land and hold it until the land price increases, as the opportunities for developing the land increase with the encroachment of urban land development.

Vacant land prices are adjusted by the weighted average of the price adjustments (in the current year) of each of the developed building types in a location. After the location of businesses and households triggers a market adjustment in land prices for each building type, these price adjustments are applied to the vacant parcels in a zone in proportion to the acreage of land containing each building type.

Database Development

The database developed for the implementation of UrbanSim contains the following elements:

- Parcel GIS database containing approximately 73,000 parcels within the Lane COG planning area
- Business establishment database for approximately 6,000 businesses within the planning area
- A household database synthesized from census tabulations by census block group and the 5 percent Public Use Microdata Sample
- GIS themes representing land development constraints, including the Urban Growth Boundary, steep slopes, wetlands, floodways, stream buffers, and utility line easements
- Zone to zone travel impedances from the travel demand model, for computing accessibility measures
- Regional control totals, specifying the overall level of population and employment growth

The parcel data used in this project originated from the Lane Council of Governments (LCOG) and was available for 1994 in ArcInfo format. The database contained assessed land and improvement value, land use code, land use plan designation, and lot size. Although it included the number of residential housing units, the database did not originally contain the square footage of nonresidential buildings. Because the model explicitly accounted for the land market components of land, structures, and occupants, the square footage data were critical for the implementation of the model. Without square footage, the basic supply information for the nonresidential components of the model was missing. LCOG had conducted a study in Eugene in 1994 for developing input data for their transportation models that included inventorying square footage on nonresidential parcels. A supplemental effort was undertaken by LCOG as part of this project to collect comparable data for nonresidential parcels in Springfield and the rest of the planning area.

Other elements of the parcel database that warrant further description include two aspects of its database design. First, ownership parcels were subdivided by LCOG into land use polygons wherever an ownership parcel contained more than one land use. The dilemma created by this design was that although the area and land use of these subparcel polygons were known, other available data such as land value, improvement value, and square footage of nonresidential space, were known or collected only at the level of the ownership parcel. This limited the ability to exploit the land use data, and for the purposes of the modeling, the ownership parcel was chosen as the basic unit of data. A second database design issue became prominent in subsequent data analysis: Wherever a building or complex of buildings, such as a mall, crossed parcel boundaries, the building or buildings were assigned to one of the overlapping parcels. This would not have been a concern if the geocoding of businesses had been entirely consistent with the assignment of buildings to parcels. But businesses were often assigned to the parcels adjacent to the one to which the building had been assigned, thus creating the appearance of one parcel with a vacant building, and an adjacent parcel with a "homeless" business.

The business establishment database maintained by LCOG is based on the State Employment Commission employer database. It has omissions regarding self-employment and proprietor establishments and, as in most states, has an additional problem in that all of the employment within a multiestablishment business may be reported at the address of the headquarters or of a single administrative office. Since 1978, however, LCOG has had an exceptional program in place to obtain and geocode business establishment data biennially and to allocate headquarters employment to individual establishment locations. In addition, the geocoding procedures at LCOG are based on the use of a Master Address File that links addresses to individual parcels. This means that address-matching the business establishment records links them to the land ownership parcels.

To develop a household database geocoded to the level of Traffic Analysis Zone (TAZ) and housing type, census data and parcel data were combined in an innovative synthesis approach. Procedures for imputing or synthesizing household data by combining geographically detailed census tabulations with sample data, such as the 5 percent Public Use Microdata, have been described and used by various researchers in the context of developing microdata for models based on microsimulation; see, for example, Clarke (1996). The techniques involved are typically based either on Iterative Proportional Fitting (IPF) or on reweighting of sample weights in a survey sample. The approach used in this project employed IPF to allocate household samples to census block groups, using marginal tabulations by block group, and to use parcel data on housing units by type by Traffic Analysis Zone, and by census block group to assign households to Traffic Analysis Zones.

The resulting database contained households and businesses stratified into groups and geocoded to combinations of Traffic Analysis Zones and building types. Information on the parcel characteristics of each of these building objects was derived from the parcel database, including the quantity of housing units and nonresidential square footage, land and improvement values, and acreage. A link to the fully disaggregated parcel database is retained both in the initial database and throughout the execution of the UrbanSim model.

Simulating Land Capacity

The model simulates urban development on the basis of scenarios consisting of the land use plan, environmental constraints, density restrictions, development costs, urban growth boundaries, transportation infrastructure (which generates accessibility measures through the travel models), and expectations of aggregate economic growth in the region. The outcomes predicted by the model include the following:

- Land, housing, and commercial real estate prices
- Quantity of real estate development by type
- Quantity of land developed for each use, and remaining vacant land
- Development density by type of development
- Quantity of development occurring on vacant land and from redevelopment
- Location of businesses and households by type
- Measures of consumer surplus by type of household and business

These appear to be sufficient measures with which to assess land capacity. The model provides a framework to undertake multiple analyses with scenarios that combine different land use and transportation policies, and to evaluate their impact on urban development and on the degree to which the goals of growth management are being achieved.

Refining the Model

Several limitations in the current model implementation have been addressed in recent updates. They include the management of large developments "in the pipeline" by allowing the user to enter large development events that the model would be unable to predict. A similar issue applies to major business events, such as the recent location of a major semiconductor plant in Eugene.

An additional refinement of the development module relates to the way it simulates development projects on large parcels. Currently, the entire parcel is assumed to develop at once and is allowed to develop into only one land use. Clearly, this is unrealistic for large parcels, for two reasons. First, large-scale developments may take multiple years to complete construction. The completion may be phased in, as with residential subdivisions or even multi-family residential development, or may be withheld from the market until completion, as with a large office project. Further, the parcel may be subdivided, with different uses on sections of the original parcel, or may be developed as mixed use within the same parcel. These refinements have now been implemented in the model.

A common concern raised about the use of parcel data is the difficulty of spatial processing of parcel data in a polygon-based GIS. This also applies to the model. The overlay operations of environmental constraints require substantial processing, and proximity calculations such as the determination of the mix of land uses within a quarter mile of each parcel would be entirely prohibitive. One solution is the use of a grid-cell conversion of the parcel data and other layers to facilitate efficient spatial processing. For this to be done without significant loss of data, the cell size for the conversion would have to be smaller than a small-lot residential parcel, or approximately 50 feet on a side. At this resolution the volume of the raster data becomes quite substantial, and other trade-offs make the benefits less clear. For example, information about the original configuration of the parcel, such as its size and geometry, may be useful in determining the feasibility of alternative development projects. The size and geometry of the original parcel will be lost upon conversion to a grid representation, unless substantial effort is made to retain a linkage between the original parcel data and its raster representation.

Clearly, additional work is needed to address not only some of these logistical questions, but also substantive issues relating to the construction of the model and its validation relative to observed trends and patterns of development. Perhaps the difficulties of working with parcel-level GIS data may be more readily addressed if its use is leveraged for planning objectives that are in demand, such as the need for land capacity analysis and for urban simulation modeling for growth management.

CONCLUDING COMMENTS

The real test of the potential value of urban modeling is whether it can be made relevant to planning practice. Such practice occurs not just at the metropolitan scale of regional transportation planning, but also at the municipal and neighborhood scale where the land policy decisions embodied in capital improvements plans, comprehensive land use plans, and other land policies are actually made. Although some of these plans may have the relatively long time horizon of 10 to 20 years, they are more frequently updated and may need monitoring on a short-term, systematic basis to assess compliance with legislative mandates related to growth management objectives.

How might an urban model such as UrbanSim actually be of value in these planning activities? One approach is to adapt the model as a general-purpose planning tool that blends the planning functions of developing and revising comprehensive land use plans, reviewing capital improvement programs, and monitoring compliance with growth management mandates. This would entail a synthesis of the functions of land monitoring and land use modeling in a way that has not previously been attempted. It would also require a multiparticipant collaboration that engages municipal, county, metropolitan, and, possibly, state agencies as partners (see Chapter 7). Certainly, the potential technical components of an integrated system to support this approach exist, such as longitudinal GIS, Internet-distributed databases and applications, and collaborative decision support systems, even though they may not have been previously integrated or applied to such domains as land capacity monitoring or urban simulation modeling.

The technical issues pertaining to networking, distributed databases, and applications to make such a vision become real are trivial, however, compared with the political and organizational obstacles that would have to be overcome. Organizational turf, proprietary approaches to data, political fragmentation, limited funding, and innumerable other bureaucratic hurdles obscure the path. Nevertheless, perhaps the technology to develop such a collaborative planning system and the political will to overcome bureaucratic impediments to its development now lie within reach.

REFERENCES

Clarke, G. P. 1996. *Microsimulation for urban and regional policy analysis.* London: Pion Limited.

Godschalk, David R., Scott A. Bollens, John S. Hekman, and Mike E. Miles. 1986. *Land supply monitoring. A guide for improving public and private urban development decisions.* Boston: Oelgeschlager, Gunn & Hain, in association with the Lincoln Institute of Land Policy.

Landis, John. 1995. Imagining land use futures: The California Urban Futures Model. *Journal of the American Planning Association* 4: 438–457.

Landis, John, and Ming Zhang. 1997. Modeling urban land use change: The next generation of the California Urban Futures Model. Presented to the National Consortium on Geographic Information and Analysis: Land Use Modeling Workshop.

McIntire, James L., and Gregory R. Easton. 1997. *The impact of urban growth boundaries on housing costs in King County.* Report prepared for the Association of Suburban Cities, King County, Wash.

Mildner, Gerard C. S., Kenneth J. Dueker, and Anthony M. Rufolo. 1996. *Impact of the urban growth boundary on metropolitan housing markets.* Portland, Oreg.: Portland State University for Urban Studies.

Portland Metro. 1997. *Urban growth report: Final draft.* Portland, Oreg.

Vrana, Ric, and Kenneth J. Dueker. 1996. *LUCAM: Tracking land use compliance and monitoring at Portland Metro.* Report to Metro, Portland, Oregon. Portland: Center for Urban Studies, Portland State University.

Waddell, Paul. 1998a. Exploiting parcel-level GIS in urban modeling. *Proceedings of the ASCE Conference on Land Use, Transportation and Air Quality: Making the Connection.* Portland, Oreg.: American Society of Civil Engineers.

———. 1998b. A behavioral simulation model for metropolitan policy analysis and planning: Residential location and housing market components of UrbanSim. Paper presented at the Eighth World Congress on Transport Research, Antwerp, Belgium.

COMMENTARY: ISSUES AND OPPORTUNITIES PRESENTED BY URBANSIM

Nancy Tosta

Paul Waddell has written a provocative chapter on urban simulation modeling. He describes a model that he has spent several years refining (UrbanSim) and how it might be used to assist in the process of land capacity monitoring. He argues that although there is increasing discussion about monitoring land supply and debate about the resolution at which this must be done, including the parcel level, there is no similar discussion devoted to assessing the level and composition of demand for urban land. He sees this as a major shortcoming of most of the existing approaches to measuring capacity. He describes the work he has done in developing demand functions by estimating consumer preferences expressed through choices made in recent moves. He examines a number of variables to generate "household bid functions" based on stratification by income group and presence of children. He continues to describe how the model utilizes parcel data to simulate the development process. The combination of a highly detailed land inventory and the ability to simulate consumer preferences and parcel-level development makes the model potentially useful to the process of land capacity monitoring.

I agree that the lack of understanding of what consumers are most likely to do diminishes a true understanding of the land supply picture in most land capacity analyses. Waddell's efforts to address this shortcoming are important. However, attempts to model such complex behavior require that certain variables be selected and analyzed, and others to be ignored. The variables that are ignored are more likely to be those that are difficult to measure, even though they may be more important. For example, although the model takes into account access to employment and travel time to the central business district, other variables, such as access to good schools and perception of neighborhood crime, are not addressed.

An additional challenge presented by the UrbanSim approach involves the requirements for data quality. Waddell outlines the problems related to acquiring parcel data that are accurate and of good quality: missing or inappropriate attributes, lack of availability in digital form, lack of connection between spatial and attribute data, limitations on data use, and out-of-date files. Relatively few jurisdictions in the nation have the depth of experience in GIS of Lane County, Oregon, the site of the UrbanSim pilot. Lane County maintains a large number of accurate data layers, and it is a jurisdiction with relatively few parcels. The intensity of data development that is required to run the model may impede its usefulness in larger jurisdictions that have not invested in accurate parcel databases. Waddell seems to express concerns of his own about the model, near the end of this chapter, with the statement,

"Clearly, additional work is needed to address not only some of these logistical questions, but also substantive issues relating to the construction of the model and its validation [relative] to observed trends and patterns of development." Aside from these practicalities, the scope and thinking in the chapter are valuable to consider in relation to what might be done someday when we catch up with the capability that technology offers.

COMMENTARY: ADVANCES IN MODELING OFFERED BY URBANSIM

Kenneth J. Dueker

This chapter explores a range of approaches to assessing land supply relative to demand. The author stresses the importance of the demand side, expressed as the composition of urban land consumers and their preferences for development types and locations. The UrbanSim model attempts to incorporate the behavioral foundation of key actors—households, businesses, developers, and governments—in simulating land use change. The structure of UrbanSim is a nested locational choice model that uses bid functions to simulate the demand of households and businesses for land and built space. Development activity is simulated by a profit calculation.

The renewed interest in urban land use models stems from pressures on metropolitan planning organizations (MPOs) to systematically model the interactions of land use, transportation, and air quality. As the author points out, models are equally needed to assess the interaction of supply and demand for urban land within the context of urban growth management.

The UrbanSim model is an important contribution. It is comprehensive and has a logical theoretical structure. Earlier models did not connect the supply and demand sides very well, tending instead to concentrate on one side to the exclusion of the other. Land capacity approaches are generally driven by supply—based on the inventory of developable lands—with inadequate consideration of demand differences. On the other hand, land use models generate demand-driven allocations of regional forecasts of population and employment to small areas based on highway accessibility. In these models, land capacity functions as a constraint to ensure that excess demand is not allocated to zones that are built out.

UrbanSim is a robust model of both the demand and supply sides at a low level of aggregation. Specifically, the parcel is important. The supply-side model operates at the parcel level, and physical limitations to development and planning restrictions are applied to parcels. Then the expected profitability of all parcels is rank-ordered. The author discusses the trade-offs of using polygons versus 50-foot grid cells in terms of efficient spatial processing to incorporate physical limitations and planning zones. Market clearing is accomplished by adjusting land prices according to the ratio of demand to supply in each zone.

The UrbanSim data units—polygons, parcels, and zones—are adequate abstractions of the real world for the purposes of monitoring land consumption. The modeling and data choices made in developing UrbanSim avoid the difficult problems of dealing with individual buildings, building permits, and individual businesses and households, along with the changes associated with each. UrbanSim demonstrates the power of choosing well the abstract units of analysis and relevant data. However, UrbanSim is not fully operational. A number of data problems and empirical estimation issues remain to be resolved.

I commend this effort not only for contributing technically to the field, but for responding to the need for a comprehensive urban simulation model. Too many visionary, long-term planning efforts are based on political or "wishful thinking" allocations of population and employment. "New Urbanism" planning efforts call for higher densities, more mixed uses, more redevelopment, and less fringe area development than the market would normally provide. Models like UrbanSim can be used to test New Urbanist concepts before committing to their implementation. UrbanSim promises to make possible realistic and accurate conditional forecasts of the future location of population and employment on which to base plans.

Conclusions

Anne Vernez Moudon and Michael Hubner

As a distinct element of urban and regional planning, land monitoring seeks at once to assess the state of the supply and capacity of buildable lands and to track land use and land development over time. Our research shows that today, the most advanced forms of land supply and capacity monitoring (LSCM) exist mainly within the arena of growth management. Various growth management programs in place nationally support land monitoring through legal mandates, methodological and technical guidance, and specific funding to cover the considerable costs of implementing and maintaining monitoring systems. LSCM is commonly integrated into programs to help reduce the impacts of urban development on both the consumption of land resources and the costs of land servicing. Evaluating urban land supply and capacity in a growth management context therefore considers both the regulation of land use and the provision of urban infrastructure as key determinants of development potential. In addressing these issues, this book exposes the "mechanics" of managing urban growth, including detailed systems of data collection, analysis and program evaluation that shape and provide feedback to the formulation of policies and regulations. By highlighting the complex procedures needed to monitor urban land, the book has sought to add an important technical dimension to an existing body of literature on growth management that primarily emphasizes public policy perspectives. Hence, this work engages the implementation of growth management in areas that are of keen interest to policymakers, planning practitioners, and scholars who seek to foster environmentally and fiscally sound urban regions.

Monitoring Land Supply with Geographic Information Systems, edited by Anne Vernez Moudon and Michael Hubner ISBN 0 471371673 © 2000 John Wiley & Sons, Inc.

The book also assesses the importance of advances in GIS technology to land information management and, correspondingly, to land monitoring practice. Planners are relying on increasingly detailed data, particularly spatial data, and have developed sophisticated approaches to land monitoring that capitalize on these new resources. Although it is clear that geographic information systems (GIS) have become the primary tool for LSCM, it is important to remember that it is but one of an array of tools for land supply monitoring, which includes permit tracking systems; spreadsheet, statistical and database software; and remote sensing. Within this context, however, GIS performs a significant role as a platform for integrating all of these tools and the data associated with them within a single spatial database framework.

Against the evolving technical backdrop, this book fills a gap in the literature on the theory and practice of land monitoring by updating the factual and critical information on LSCM practices across the country and by proposing new frameworks for understanding these practices. Overall, the focus remains on the technical, methodological, and organizational aspects of LSCM and GIS, with correspondingly few critical references made to policy considerations or to the politics of growth management.

In closing this book, we believe it is important to reiterate the impressive potential of GIS to meet both present and future needs of land monitoring. The proliferation of GIS in local government, in particular, promises to support both the improvements to and the expansions of current land monitoring applications. Further development of parcel-based geographic information systems (PBGIS) is seemingly inevitable as they are applied to enhance, streamline, and make more efficient a wide range of standard local government functions. We find that the implementation of these systems offers at least four significant opportunities to support and improve current practices of LSCM.

First, PBGIS represent land in a much more spatially detailed way than previously feasible. Spatial patterns of land use change and change potential can now be readily and accurately identified for small planning areas, such as urban centers or transit-oriented development nodes and corridors. Second, PBGIS may be linked to transactional data, such as permits for land subdivision and building construction and occupancy. This allows the integration of the day-to-day management of land use change with long-term comprehensive planning and plan evaluation. Further, the ability to monitor land development processes at the parcel level means that the arena of private real estate and development may be considered in, and perhaps reconciled with, traditionally separate, long-range, and geographically coarser regional land planning perspectives. Third, spatially detailed inventories of land use and land supply can be tracked over time to compile a longitudinal record. This capability is essential to identify trends that then serve to inform analysis assumptions about future land absorption and development patterns. Fourth, parcel databases may now be constructed or assembled for entire urban regions. This capability is critical because of the regional scope of real estate markets and land development practices. Not surprisingly, therefore, the benefits of parcel and even subparcel data are increasingly acknowledged and pursued even in the context of regional planning.

PBGIS thus make possible entirely new ways of describing land use and land use change and, by inference, new ways to analyze the interactions between public policies and regulations governing land use and private processes of land use change. As, at this writing, GIS and land monitoring systems remain short of realizing fully these opportunities, this book deals with ways to improve technical practice. Several different themes emerge.

First, various of the dimensions of land use and land supply remain difficult to define and measure, both geographically and temporally. One prominent example concerns multiple and mixed uses, which do not readily lend themselves either to mapping or to tracking over time. Another involves the analysis of infrastructure service levels, which are not easily translated into quantities and precise locations of serviced land. In the environmental realm, the accurate and precise mapping of factors that constrain development presents problems as well, especially at levels of resolution adequate to analyze the suitability of individual parcels.

Second, ongoing debate surrounds the issue of appropriate geographic units of data measurement, representation, and analysis. This book has focused primarily on the potential and limitations of the parcel unit. Although parcels are crucial to land supply and capacity at all points in the development process and for all types of land, not all data are appropriately parcel data. Chapter 3 argues for the utilization of mixed spatial units to monitor land in various stages of development. Examples from the field, such as the "industrial concentration" approach described in Chapter 5, and the Portland Metro land monitoring program, illustrate a range of creative approaches to building and analyzing databases—both parcel and nonparcel—on land supply. The key for the future will be monitoring the relationships between parcel data and data associated with other geographic units as discussed in Chapter 6.

Notably, PBGIS are gaining importance for regional planning, where coarser levels of data than the ownership parcel have predominated for some time. However, there are particular barriers to implementing PBGIS at the regional level. Parcel files continue to be unwieldy from a computational point of view, especially for processing spatial analyses over large areas. More important, parcel coverages simply do not yet exist as reliable and compatible multijurisdictional data sets for most urban regions. For these reasons, alternative approaches, such as aerial photographs and land supply data aggregated to analysis zones, continue to dominate regional monitoring programs.

Third, tracking land supply changes through time poses significant problems. Chapter 3 describes a methodological and database framework for maintaining a perpetual inventory of supply and capacity, and Chapter 6 explores the representation of temporal aspects of land information in database systems. As highlighted in these and other chapters, despite successful implementation of longitudinal monitoring in select cases, there remain significant technical and organizational barriers to constructing ongoing land inventories and time-sensitive analysis approaches.

Fourth, local government organizational structures often stand in the way of building and maintaining land information systems adequate to the tasks of LSCM—as discussed in Chapter 7. The scope and complexity of the data requirements of LCSM dictate that planners rely on their ability to leverage development of and access to data from many units within local government. However, the organizational support for such data sharing is uneven at best, especially as multidepartmental and multijurisdictional efforts work against the inertia that is common among public agencies. On one hand, decentralized responsibilities for land monitoring across and within jurisdictions hold some promise for negotiating these difficulties. On the other hand, the creation of centralized institutionally separate agencies to collect, maintain, and analyze data offers a more coordinated top-down approach. However, centralization requires local governments to agree to pool their technical and financial resources in the service of a shared mission, which may be difficult outside a strong regional governance structure.

Overall, evidence from the case studies suggests that building GIS databases for LSCM is a long-term project, which, because of ongoing technological change, requires flexibility and adaptation as well as foresight, tenacity, and considerable financial support. Because these are not common characteristics of the public sector, it appears likely that over the long run local jurisdictions will build land monitoring systems through gradual upgrade and not through wholesale replacement. This, in turn, suggests that decentralized approaches to land monitoring will persist in the future.

The initial impetus for this book lay in advancing substantive approaches to urban planning and growth management and the development of urban and suburban lands. The book aims to foster progress in the understanding of methodological, technical, and organizational issues that shape land supply and capacity monitoring as a means toward these ends. This being achieved, several arenas for future work emerge.

The accumulation of longitudinal data on land development will likely have the greatest effect in advancing approaches to urban land planning. Planners will need to develop appropriate uses of these data to effectively address the future. As a prime example, the question of how closely to base assumptions about future capacity on information gathered from short-term development trends reveals tensions between public and private stakeholders. Generally, growth management planning tries to achieve a balance between accommodating market forces that may be evident in existing development patterns and taking a more active role in influencing the distribution and character of demand and resulting development. Yet these two orientations imply two different approaches to assessing future land use potential and require different analytical procedures. Careful analyses of longitudinal data collected through consistent and ongoing land monitoring will help to balance the consideration of development trends with the formulation of effective policy.

Longitudinal data may be applied to comparative research in several areas. Such data will be useful to assess the effectiveness of various land monitoring approaches. Measures of effectiveness should address how well LSCM programs have assessed the availability and development potential of local and regional land resources in relation to the requirements of specific growth management frameworks. When land monitoring agencies have sufficient data to adjust their practices appropriately, it will become possible to compare and assess long- versus short-term supply and capacity estimates. Specifically, it will be feasible to assess the accuracy of estimates established by different methods. Evaluations can also serve to identify the strengths and weaknesses of centralized regional monitoring systems, such as Portland Metro's, versus decentralized ones, such as the central Puget Sound region's. As well, opportunities will exist to compare various alternative tools used to regulate land supply such as urban growth boundaries versus adequate public facilities ordinances versus phased extension of infrastructure service areas.

Furthermore, effective land monitoring both requires and provides advanced capabilities in the forecasting of amounts and characteristics of the demand for land. As described in Chapter 8, urban simulation models can address two problematic aspects of current monitoring practice. The use of models offers an alternative to (1) the current treatment of the land supply as a homogeneous container for growth, undifferentiated by either location or development type, and (2) the use of static assumptions about supply and demand interactions—which are inherently dynamic—over long time horizons. Disaggregated computer models of urban development are an important avenue for the future evolution of LSCM, especially as they can help to forecast, more effectively than current land capacity analyses, where, when, and how buildable land will or can be consumed.

In all of these areas, academic research will be required to provide better theoretical grounding for practice. Such research is likely to flourish in the wake of the rapid diffusion of GIS and improved LSCM databases. Key areas include real estate economics, alternative urban form and development patterns at both the regional and local scales, interactions between transportation and land use, and the effective use of GIS in public agency settings.

Last, but not least, research must turn to the actual results of LSCM and study the amounts, types, and distributions of the developable land supply within urbanized regions. So far, most monitoring programs have found there to be ample land supply and capacity relative to anticipated demand. This is particularly interesting in that many studies have used assumptions which limit the definition of buildable lands through the application of discounts for market availability and infrastructure servicing. Similarly, estimates of capacity have paid lip service to contemporary suburban development patterns, and have assumed the continuation of low-yield development practices such as residential under-build and preferences for large lots for industrial uses.

Detailed assessments of reported land supply surpluses seem critical and timely. A range of commonly overlooked factors, such as location preferences and amenity considerations that shape land markets, will need to be considered. Such efforts will likely encounter contradictory reactions in the public arena. One is the strong resistance to change in American cities and to increasing development density in areas deemed already urbanized (with the exception of downtowns). This resistance reflects discontent among existing residents as well as opposition from the development industry, which favors large tracts of land at the urban fringe. A second reaction is the increasingly strong political resistance to expanding urban growth areas, which has persisted in the face of considerable demand pressures and has fueled resurgent efforts to redirect growth within existing centers. Results emerging from monitoring programs already suggest that urban development may selectively and unexpectedly be turning inward on the existing urban fabric. Portland Metro, for instance, found that 29 percent of new residential building permits occurred on lots that had not been classified as vacant in earlier inventories (meaning, most likely, that the private sector found demand-driven opportunities in this land). Parcel-based LSCM can support planners in capitalizing on this apparent trend. PBGIS may be used to identify the opportunities that are potentially missed in fully utilizing the supply of land already serviced, as well as to provide a more precise picture of the amounts and types of lands skipped over by development. This information can help planners to become proactive and make such lands attractive to developers.

LSCM holds the potential to enhance planners' understanding of land consumption patterns and development practices. Improved understanding will lead to improved means of achieving desired forms and patterns of land development, efficient provision of infrastructure, and protection of the environment. Urban land monitoring will clearly continue to evolve, especially as an increasing number of local, regional, and state governments question the wisdom of perpetuating current land development practices that produce low-density sprawl. Other city-regions will join the cases noted in this book in taking steps to manage the development of urban land resources and to monitor and evaluate the supply of land to meet both public- and private-sector needs. Although, for many small cities, land monitoring is likely to remain sporadic and relatively narrow in its focus, the availability of GIS, and PBGIS in particular, will enable planners and policymakers at many levels of government to carry out more detailed and continuous applications of land monitoring in the service of both strategic planning and development management. Opportunities abound to explore the relationship between the land development objectives of growth management and broader goals to foster livable urban environments, sustainable economic development, and the preservation of natural and human communities.

Appendix A

Survey of Land Supply Monitoring Practice[1]

This appendix reviews the findings of a national survey of local and regional planning agencies, synthesizing information on the state of the practice in land supply and capacity monitoring in 38 of the jurisdictions contacted for the project for the Lincoln Institute of Land Policy. The first section describes data collection methods and gives a general overview of the cases selected for review. Subsequent sections address the following three dimensions of practice: (1) strategic planning applications of land monitoring systems, (2) database development, and (3) land supply and capacity analysis methods. Illustrative examples from practice, as well as summary tables, accompany the text.

SURVEY OVERVIEW

Data Collection

Data on land supply monitoring systems in use were collected from October 1997 to March 1998. In the tradition of exploratory research, the survey

---

[1] This analytical summary is derived from Michael Hubner, "Urban Land Supply Monitoring: A Critical Review of Current Practices in Measurement, Analysis, and Application" (master's thesis, University of Washington, 1999).

Monitoring Land Supply with Geographic Information Systems, edited by Anne Vernez Moudon and Michael Hubner ISBN 0 471371673 © 2000 John Wiley & Sons, Inc.

approach emphasized comprehensiveness and flexibility, encompassing both the reach of current practices and their limitations.

A literature review and interviews with academic and practicing planners guided the purposive sampling of cases that met, either in whole or in part, search criteria as follows. First, the survey followed up on land monitoring case studies reviewed by Godschalk et al. (1986) or cited elsewhere in the published literature.[2] Second, jurisdictions with growth management programs described in the planning literature were contacted. Third, several jurisdictions characterized by interviewees as having well-established or state-of-the-art land information systems (especially incorporating parcel-based geographic information systems) were contacted. Finally, jurisdictions in Washington State received special emphasis for two reasons. In compliance with the state's Growth Management Act (GMA) (1990), local governments throughout the state recently conducted land supply and capacity analyses to establish their urban growth areas (UGAs). Furthermore, recent amendments to the GMA (1997) introduced new requirements mandating detailed land monitoring within six urban counties in the state.

Data on selected cases were collected from a variety of primary and secondary sources, including phone interviews, plan documents (e.g., plans, policy documents, technical reports on land supply and capacity and related topics, information system descriptions), published articles, and government web sites. Interviews with agency staff—primarily planners, geographic information system (GIS) analysts, and private consultants—served as primary sources of information. As the survey progressed, land monitoring systems were found to vary considerably in their level of complexity, degree of formalization, and completeness of documentation. Some of the more well established land supply monitoring programs had recently developed enhanced GIS data resources and revised methodological approaches to monitoring. For these reasons, the quality and quantity of data collected on the various cases differed significantly.

Interviews with planning professionals in the field focused on recent and ongoing land monitoring activities, information systems in place, land use planning contexts, and a range of case-specific issues. Survey interviews followed a flexible protocol and varied in scope and focus. Questions addressed included the following:

- Types of land supply data collected—emphasizing GIS, parcel files, permit tracking, environmental data, and market-related information
- Data sources—including in-house data development (typically by planning departments) and inter- and intra-jurisdiction data sharing

---

[2] David R. Godschalk, Scott A. Bollens, John S. Hekman, and Mike E. Miles. 1986. *Land supply monitoring: A guide for improving public and private urban development decisions.* Boston: Oelgeschlager, Gunn & Hain, in association with the Lincoln Institute of Land Policy.

- Geographic units of data collection, representation, analysis, and reporting
- Geographic areas for various monitoring and evaluation activities
- Methodologies for evaluating buildable land supply and development capacity
- Types of land development potential considered in evaluating land supply (e.g., vacant land, infill sites, and redevelopable properties)
- Types of land use potential considered in evaluating land supply (e.g., residential, industrial, commercial)
- Constraints considered in evaluating land development or redevelopment potential (e.g., physical, infrastructure, environmental, market-related)
- Measures of overall amount and composition of expected future demand for land and built space
- Role of GIS (particularly parcel-based) as a land supply monitoring and analysis tool
- Recent or periodic reports, inventories, studies, or plans generated by the land monitoring program
- Applications of land supply information for planning and land management purposes

Study Cases: Selection and Major Characteristics

The 38 cases selected and reviewed represent the full range of practice in land supply monitoring nationally. The cases include 13 municipalities, 16 counties (9 include central cities, 7 encompass only suburban areas), and 9 regional planning agencies (8 urban regions, 1 nonurban region). The jurisdictions range in size and type—large and small central cities, urban and suburban counties, and multicounty urban regions—and are located in all major regions of the United States. Table A.1 lists the cases by jurisdiction type.

All 38 jurisdictions had implemented computerized land information systems with fully or partially integrated GIS capabilities. Thirty-three of the systems included parcels as a base data layer (many of these parcel coverages had only recently been completed). The use of GIS and land information systems (LIS) more broadly, by jurisdictions, extended beyond land supply monitoring to include other long-term planning activities, short-term planning, and ongoing land management. Automated permit tracking systems were employed by a number of agencies, although these were often not well integrated with GIS or other land information databases. Agencies reported utilizing a range of data types in their monitoring efforts—including mapped, tabular, and image data—that were compiled from a variety of data sources. Land supply database management entailed both intra- and interjurisdiction data sharing.

Within urban regions, monitoring responsibilities were commonly distributed among local jurisdictions. For example, the four cities in Washington State

TABLE A.1 Selected Cases for Review

Type of Jurisdiction	Name of Jurisdiction
Municipality	Anchorage, Ala.; Boston, Mass.; Fitchburg, Mass.; Honolulu, Hawaii; Kansas City, Mo.; Kent, Wash.; Newark, N.J.; Ontario, Calif.; Redmond, Wash.; San Jose, Calif.; Scottsdale, Ariz.; Seattle, Wash.; Shoreline, Wash.
County	Boulder Co., Colo.; Clark Co., Nev.; Dade Co., Fla.; DuPage Co., Ill.; Johnson Co., Mo.; King Co., Wash.; Lee Co., Fla.; Loudoun Co., Va.; Marin Co., Calif.; Marion Co., Oreg.; Montgomery Co., Md.; Orange Co., Fla.; Snohomish Co., Wash.; Suffolk Co., N.Y.; Wake Co., N.C.; Washoe Co., Nev.
Regional government or planning agency	Association of Bay Area Governments (ABAG), Calif.; Cape Cod Commission, Mass.; Fargo-Moorhead Council of Governments, N.D.; Lane Council of Governments (LCOG), Oreg.; Metropolitan Council of Governments, Minn.; Northeastern Illinois Planning Commission (NIPC), Ill.; Portland Metro, Oreg.; Puget Sound Regional Council (PSRC), Wash.; San Diego Association of Governments (SANDAG), Calif.

surveyed generated land supply and capacity reports that were subsequently incorporated into countywide monitoring programs. Seven counties participated in regional land monitoring, conducting analyses at the local level for incorporation into regional efforts. In turn, several regional and county agencies provided guidance for constituent jurisdictions in the form of methodology guidelines, reporting requirements, data provision, and direct assistance in conducting analyses.

Levels of future demand for land were most commonly derived from regional long-range demographic and economic forecasts for planning periods of 5 to 20 years or more. Local demand was often established through the allocation of regional forecasts to small areas. Such allocation occasionally employed computer modeling techniques to simulate future growth patterns, but more commonly relied on simple projections of local trends in population and employment, modified by targets for future growth established within comprehensive plans. In addition to expected demand for a total amount of urban land, the composition of the demand also served as an input to monitoring analyses. Jurisdictions typically used data on 2- to 10-year trends to predict future compositional factors such as housing type split, average lot size, square feet of land per building type, and floor area per employee.

The types of geographic areas over which jurisdictions inventoried and tracked land supply varied considerably. Areas being monitored and evaluated ranged in size and type, encompassing land demarcated by or referenced to political boundaries, land use and activity patterns, planning and land management zones, and natural features. Table A.2 lists examples of monitoring areas by type.

The geographic extent of land monitored by specific agencies closely reflected planning frameworks (especially where monitoring objectives were

TABLE A.2   Geographic Areas for Land Supply Monitoring

Reference Type	Monitoring and Evaluation Areas
Political units	Multicounty regions, local jurisdictions (county, city, and town), annexation areas
Use and activity areas	Urbanized regions, large subdivisions/development projects, transportation corridors and functional nodes
Planning and land management zones	Urban growth areas (UGAs), district and neighborhood planning subareas, plan-designated centers or nodes, zoning/planned land use districts, transportation analysis zones/forecast analysis zones (TAZs/FAZs), infrastructure service-sheds
Natural features	Watersheds, natural hazards risk zones

dictated by policy and administrative requirements) and jurisdiction types. A majority of the jurisdictions monitored land supply over their entire land areas. Counties commonly restricted their efforts to unincorporated areas over which they had sole planning jurisdiction, but often performed (or coordinated) monitoring for the cities within their boundaries as well. Entire UGAs and other urban containment areas—usually administered by counties and regional agencies—defined the primary land supply focus for nine of the agencies surveyed (four of the cities fell within UGAs and planned within urban containment guidelines). Growth management policy also dictated other smaller-scale monitoring areas, such as centers designated for concentrated growth (e.g., Urban Villages) and annexation areas (where these were closely tied to urban area expansion). In concordance with the various scales at which planning occurred, land supply monitoring within single jurisdictions frequently entailed multiple geographic scales, addressing supply and capacity areawide, within planning subareas, and for areas of focused concern (typically related to the environment or transportation).

Among local government agencies, planning departments most commonly reported taking the lead responsibility for land supply monitoring. Information technology and research departments, which are more recent additions to local government bureaucracies, also played a significant or lead monitoring role in several cases (e.g., Portland Metro's Data Resource Center, Montgomery County's Research Center). Private consultants and advisory groups (including public agency, private industry, and community representatives) also contributed to the design and implementation of land supply monitoring systems and methodologies and provided technical assistance and/or specialized knowledge of land and real estate markets.

APPLICATIONS

The survey revealed that since the mid-1980s land monitoring practices had increased in complexity and scope. For example, land monitoring was applied not only to the design and implementation of urban growth boundaries, but also to planning for urban infrastructure and services, housing, economic development, environmental impact mitigation, and even natural hazards. Among the jurisdictions surveyed, the most common policy arena for monitoring land supply was growth management. Applications related to growth management addressed three primary requirements: (1) balancing supply and demand within existing and proposed urban sprawl containment areas, (2) evaluating development potential within small areas designated for concentrated growth, and (3) monitoring trends of development and land consumption for a variety of geographical areas relevant to growth planning.

Examples of both plan making and plan implementation, as well as more exploratory planning activities (such as identifying areas for targeted interventions, regulations, incentives, and investments to achieve desired land development), emerged from the survey. Table A.3 lists the major types of reported applications, along with the total number of occurrences of each. Commonly, land monitoring systems were used for more than one application type within a single jurisdiction.

Comprehensive supply and capacity. The most often reported land monitoring approach (17 cases out of 38) entailed conducting single or repeated areawide snapshots of land supply and capacity for comparison with medium- to long-term growth forecasts. Among the cases reviewed, land monitoring contributed to the designation of UGAs, as well as subsequent assessments of the adequacy of UGA capacities with changes in prevailing development and demand over time. Most cases of comprehensive monitoring addressed infill or redevelopment potential in addition to vacant land capacity. Time horizons ranged from 5 to 20 years, and various techniques were used to account for land availability and development feasibility. Examples of comprehensive monitoring applications included the following:

- *Portland Metro.* Ongoing periodic monitoring of land capacity within the three-county UGA. The latest comprehensive analysis of residential and employment land capacities (1997) incorporated new 2040 growth concept assumptions, including increased densities along corridors and within nodes. (See Appendix B.)
- *Seattle.* Background studies of land capacity supported the development of the city's comprehensive plan (that met state growth management requirements). (See Appendix B.)

Focused monitoring and analysis. Among the cases reviewed, 13 projects were dedicated to assessing one aspect of the overall land supply. Several focused on specific land supply types including inventories of vacant lands (particularly at the urban fringe) and, increasingly, lands with infill and redevelopment potential. Concerned with potential shortages of land within UGAs, these jurisdictions reported seeking out new "sources" of capacity—either

TABLE A.3 Land Supply Monitoring Applications

Category (Number of Examples in Survey)	Specific Applications
Comprehensive supply and capacity applications (17)	UGA/urban service area (designation and administration) Jurisdiction-wide inventories and analyses Analyses of comprehensive land use plan alternatives Land developability/suitability database development
Focused monitoring and analysis (13)	Warehouse/office/industrial/residential/commercial supply and capacity analyses Employment supply and capacity analyses Housing inventories Vacant lands surveys Redevelopment/infill potential studies
Small-area monitoring and analysis (7)	Subarea and neighborhood planning Analyses of development capacity within transportation nodes/corridors, employment/activity centers, and plan designated growth centers
Urban area expansion studies and monitoring (6)	Annexation studies Staged growth areas planning Evaluation of lands for potential UGA expansion/urban reserve areas
Tracking land supply and development trends (11)	Tracking absorption rates within large subdivisions Plan performance indicators Monitoring regional/subregional growth and land development trends Monitoring developments of regional impact (DRIs)
Forecasting, growth allocation, and modeling (7)	Subregional population and employment forecasting/allocation Areawide population projections Land use modeling
Infrastructure planning (11)	Infrastructure demand modeling General infrastructure and facilities planning Traffic impact studies Implementing adequate public facilities ordinances (APFOs) and concurrency requirements Drainage modeling
Environmental planning (5)	Assessments of potential watershed/water quality impacts of development Natural hazards mitigation
Economic development (4)	Marketing properties for private sector property investment/development Inventories of city-owned and abandoned properties Brownfields redevelopment

"discovered" through closer, more specialized monitoring of land supply or "created" through zoning changes and incentives to developers. Exemplifying focused monitoring, Lane Council of Governments conducted a study of redevelopment within the regional UGA that explored new methods for identifying parcels with future potential to accommodate additional growth. The findings—drawing on historical data, comparative approaches, and exploratory analyses—provided information for both ongoing UGA capacity monitoring and potential strategies for encouraging land use changes in support of the regional transportation strategy. (See Appendix B.)

A second type of focused monitoring and analysis identified in the survey addressed supply and capacity for specific types of land uses. A number of cases, for example, emphasized industrial land supply as a major planning and land monitoring issue. Planning constraints on the supply of industrial lands were seen to threaten regional economic development. In addition, economic restructuring had generated concerns about the ability of planning to adjust to the changing location needs of firms as well as apparent trends to convert industrial lands to nonindustrial uses. Two recent efforts by the Cape Cod Commission (see Appendix B) and the Puget Sound Regional Council (see Chapter 5) entailed regional inventories of industrial lands. Finally, commercial development and redevelopment potential in inner-ring suburbs was another area where economic development and land planning converged in regard to questions of land supply adequacy—as in the case of Montgomery County (see Chapter 4 and Appendix B).

Small-area monitoring and analysis. Seven of the cases reviewed conducted land supply analyses for single or scattered subareas within jurisdictions. Two primary planning purposes were in play here. One was neighborhood or small-area comprehensive planning, using land supply analysis to compare land use plan and zoning alternatives. A second purpose concerned areas targeted for concentrated growth—transit-oriented development areas and Urban Villages—where capacity was typically linked with density and land-use-mix targets for these areas. Overall, small-area monitoring dealt with already built-up urban areas, with a substantive focus on infill and redevelopment. Examples included the following:

- The Lane Council of Governments analyzed development potential within 39 Nodal Development Areas throughout the metropolitan area, which were designated for concentrated land use and transit development. (See Appendix B.)
- Seattle analyzed land supply and capacity for neighborhood-based Urban Villages, designated to accommodate much of the city's growth over a 20-year planning period and targeted for strategic rezones by neighborhood plans. (See Appendix B.)
- Wake County inventoried land supply within Activity Centers designated as part of the county transportation strategy.

- Loudoun County conducted a comparative study of the land supply implications of the land use plan alternatives for a largely undeveloped county subarea.

Urban area expansion. Six of the jurisdictions surveyed monitored lands outside existing urban containment areas, especially those planned for future urban expansion. These areas included UGA reserves, staged urban service areas, and annexation lands. The areas were assessed for their capacity to accommodate population and employment increases. Other concerns included infrastructure servicing costs (a function of future land development patterns) and potential constraints on urban density imposed by rural residential uses and subdivisions. Examples of urban area expansion monitoring included the following:

- Ontario and Boulder County conducted land capacity analyses for proposed annexations. Planners in these two jurisdictions described annexation functioning as a de facto urban growth area expansion mechanism.
- The Metropolitan Council (Minneapolis–St. Paul) analyzed land supply within Urban Staging Areas, designated for urban infrastructure expansion over five-year increments through 2020. The Council also monitored development for a post-2020 urban "holding zone," delineated for potential further service extensions into currently undeveloped areas. (See Appendix B.)

Monitoring land supply and development trends. Eleven cases included dedicated monitoring components that tracked land supply and development over time, either to establish longitudinal trends within areas of concern for strategic regional planning (e.g., UGAs, large subdivided areas) or to inform areawide land policy more generally. Trend data generated through such work was used to improve analysis assumptions. The following are examples of such efforts:

- The King County "Benchmarks" program tracked land use and growth data covering proportions of new development within UGAs and designated Urban Centers, redevelopment rates, ratios of achieved-to-allowed densities, and other measures.
- Metropolitan Council: An annual report of "Performance Indicators" included rates of farmland conversion, increases in densities within urban service areas, and other measures. (See Appendix B.)
- Lee County tracked build-out rates within large speculative subdivisions (of 300,000+ lots).

Forecasting, growth allocation, modeling. In seven cases, land supply data contributed directly to forecasts of population and employment distributions.

Developable land was considered to function either to limit growth (if in short supply) or to attract growth (if in plentiful and relatively cheap supply). Subregional growth allocations—made with and without the aid of computerized land use and transportation models—were typically carried out by regional planning agencies in collaboration with local jurisdictions. Examples of such efforts included the following:

- Honolulu and the San Diego Association of Governments (SANDAG): Each of these jurisdictions combined land supply GIS data with computer simulation models of disaggregated land use change. Honolulu utilized a dynamic parcel-level model. SANDAG used a spatially detailed gravity model that was sensitive to land use and capacity factors (see Appendix B).
- Northeastern Illinois Planning Commission conducted a regional allocation of population and employment to individual jurisdictions, which was partly sensitive to capacity constraints measured and analyzed at the local level.

Infrastructure planning. Eleven jurisdictions reported applications of land supply analysis techniques for capital facilities and transportation planning. Land capacity analysis improved projections of future subarea demand for transportation, sewer, water, and storm water infrastructure. In several cases, growth management mandates regarding, for example, concurrency, drove the infrastructure focus of associated land monitoring. Planners surveyed reported general interest in monitoring the interaction between transportation and built form, residential densities, and land use. The following are examples of infrastructure planning applications:

- Dade County monitored new development for comparison with infrastructure data in order to ensure compliance with State of Florida concurrency requirements.
- Scottsdale's storm water systems and transportation demand modeling used information on parcel-level development capacity.

Environmental planning. Five cases involved land monitoring tools used to provide build-out forecasts to assess potential environmental impacts. Water resources, including both surface water runoff and protection of critical groundwater recharge areas, were a primary consideration. Impervious surfaces were also monitored, derived either from aerial imagery or from land records. In a couple of cases, capacity estimates also informed natural hazard management and mitigation, focusing on current and potential development in areas at risk of wildfire, flooding, or storm damage. The Cape Cod Commission (CCC) provided the most elaborate combination of land supply and environmental impact analyses. The CCC conducted two subregional build-out studies to provide information for a series of carrying capacity analyses that addressed water quality, critical natural areas, and "human" habitat concerns such as the integrity of rural landscapes (see Appendix B).

Economic development. Four cases highlighted the use of GIS as a tool for land monitoring in the service of local economic development. These efforts explored infill and redevelopment opportunities within economically distressed or stagnant areas. Publicly owned and abandoned lands were inventoried and evaluated for their suitability to attract reinvestment in the form of land development. One municipality, Newark, New Jersey, analyzed redevelopment potential on vacant lots and surface parking areas within its jurisdiction. In other cases, land monitoring assisted in targeting neighborhoods for grant assistance based on land use and development criteria.

Several secondary applications of land supply monitoring systems were notable among the surveyed cases: (1) current planning and land management, (2) monitoring housing and land market conditions, and (3) private-sector uses.

Current planning and land management. Many of the jurisdictions surveyed reported using land monitoring systems for current or short-range planning and plan implementation. Examples included administration of transfer of development rights programs (measuring capacity of potential donor parcels and receiving parcels), analysis of small-area rezones to determine impacts on capacity, and the use of permit tracking systems for current planning purposes.

Direct monitoring of land and housing markets. Several cases entailed monitoring land or housing prices directly. In all such cases data were captured on an aggregate basis (e.g., entire UGA, jurisdiction, or district), and in no case were market data linked explicitly to land supply monitoring systems. Portland Metro's annual Performance Indicators, for example, entailed market data collection that included housing prices (sales price for single-family houses [new vs. all homes], average rents for multifamily housing [new or recent vs. all rental units]), and land prices (primarily vacant lots). Metro also attempted to predict UGA impacts on housing markets over time. In another example, Anchorage also produced an annual report of real estate data for the urban region, covering sales prices, vacancies, housing stock, rental amounts, units on the market, permits and new units, and housing and land value indices.

Private-sector utilization of information from public land supply monitoring systems. Interviews with both public agency planners and private consultants revealed little use of public land monitoring systems by the private sector. One notable exception was the reported use of on-line site search engines linked to GIS. In Ontario, California, planners reported high levels of use by, and enthusiastic response from, the private sector, following the implementation of an on-line tool for querying the city's parcel database for potentially developable or redevelopable properties.

DATABASE DEVELOPMENT

The cases surveyed utilized a wide variety of land supply data, derived from multiple collection methods and sources. Databases ranged widely in comprehensiveness, with some of the long-established systems (e.g., Portland Metro) maintaining numerous data sets (including multiple GIS layers). Many land information systems (LIS), however, had not progressed much beyond a primary focus on data collection and base mapping. Database development work cited in the survey focused primarily on several areas: (1) constructing GIS layers for parcels, administrative boundaries, and environmental features, (2) developing automated permit tracking systems, (3) establishing standardized address formats to link various site-level data to parcel land records, (4) improving or updating inventories of existing land use, and (5) resolving inconsistencies between data sets (e.g., planned land use vs. zoning). Because they were occupied with technical database development, many agencies reported having delayed or abandoned analytical activities because of inadequate data resources and technical staff time.

Data Categories and Units of Representation

Table A.4 summarizes data items (by type), collection methods, and secondary sources. No one monitoring system collected all of the data types listed. However, data in all categories were usually present within any one agency's land supply database.

Core data commonly reported to be collected or utilized for land monitoring included existing land use, building square footage, number of dwelling units, parcel size, zoning, planned land use, ownership, valuation (land and improvements), transportation infrastructure, floodplains, wetlands, and slopes. In some cases local conditions and policy requirements dictated more ambitious land supply databases. The most extensive local or regional databases had received consistent financial and staffing support for a number of years and were well integrated into strategic long-range planning programs. Examples of well-developed multi-purpose LIS/GIS included the following:

- *Portland Metro.* The Regional Land Information System (RLIS) is a comprehensive multi-purpose GIS with more than 100 data layers, integrating parcel data, orthophotography, and permit tracking. (See Appendix B.)
- *Montgomery County.* MC:MAPS is an interagency information system with a recently completed parcel GIS layer and long-standing procedures for tracking pipeline development (maintained as separate GIS layers). (See Appendix B.)
- *San Diego Association of Governments.* SanGIS is an independent public agency that maintains more than 200 layers of GIS data, including parcels, land use plans, remote sensing imagery, and environmental features. (See Appendix B.)

A number of agencies reported efforts to capture or recover longitudinal data. However, planners reported persistent organizational and technical bar-

TABLE A.4 Data Types, Collection Methods, and Sources

Data Category	Data Type
Land use	Existing land use, secondary land use, past land uses, adjacent land uses, Standard Industrial Classification (SIC) code, employment on-site, occupancy
Physical development	Building square footage, number of housing units, parcel size, building footprint, floors/height of structure, year built, condition, floor area ratio (FAR)
Administrative status and actions	Permit status (construction and occupancy), subdivisions (approved and in process), zoning (including variances and overlays), building code requirements, planned land use, impact fee areas, agricultural reserves
Ownership, land rights, and markets	Owner name and type (public, private, church, etc.), easements, rights-of-way, valuation (land and improvements), tax status, property sales records
Infrastructure and facilities	Road type and capacity; traffic counts; street frontage; transit stations; districts, service-sheds, or levels of service for water, sewer, fiber-optic cable, and schools
Environmental and physical characteristics	Floodplains and floodways, soils, wetlands, impervious surfaces, natural habitat, steep slopes, aquifer recharge areas, vegetation, geological hazards, airport noise contours, streams and water bodies
PRIMARY DATA COLLECTION	Aerial photography, satellite imagery, windshield surveys and field methods, records of public actions and transactions, landowner surveys, digitization of paper maps, data verification from local officials
SECONDARY DATA SOURCES	Assessors, within-jurisdiction departments (e.g., building, engineering, transportation, etc.), other local and regional agencies, private data providers, National Wetlands Inventory, Federal Emergency Management Agency, state employment securities offices, newspapers and other media, private developers and real estate agents, university researchers, Multiple Listing Services

riers in this area, including ongoing overriding emphases within planning departments on assembling data on present conditions, as well as insufficient infrastructure for linking permit tracking systems to land supply databases. The following are examples of longitudinal data capture and utilization:

- Kansas City recovered subdivision records from 1985 to the present. City planners reported a 65 percent recovery rate for these data.
- Fitchburg recovered 15 years of subdivision records for systematic study of factors influencing under-build, which provided an informed basis for subsequently applying a land capacity discount to projected new development.

The agencies surveyed employed various geographic units of data collection, representation, and land supply and capacity analysis. Table A.5 lists data and analysis units by type, with examples from practice.

The parcel was the most common unit of geographic representation, with 33 programs utilizing it as a primary spatial unit for land supply monitoring (although parcels did not in all cases constitute fully digitized GIS layers). Several systems tracked land supply and capacity for units smaller than the parcel. More commonly, units of generally lower spatial resolution were used in addition, or as an alternative, to parcels. Such zonal geographies were commonly derived from policy requirements, land use surveys, census and analysis zone boundaries, or GIS overlays of multiple areas and units. Almost all of the cases employed mixed geographic units, with both database linkages and GIS overlays used to integrate disparate data types and geographic units.

TABLE A.5 Units of Data Collection, Representation, and Analysis

Geographic Units of Data Collection and Analysis	Examples from Practice
Parcels (represented as points or polygons in GIS; nonspatially in land records files)	Anchorage (polygons) Snohomish County (geocoded parcel points) Redmond (tabular parcel records)
Areas of contiguous land use, ownership status, or plan designation	Portland Metro (vacant land areas) LCOG ("sites" defined by contiguous ownership) PSRC ("industrial concentrations" defined as contiguous areas >25 acres planned for industrial uses)
Census blocks/block groups/tracts	CCC (block groups and tracts)
Units derived from GIS overlay of multiple geographic layers	SANDAG (Master Geographic Reference Areas, generated through overlay of census, planning, jurisdiction, and zip code boundaries)
Developments of regional impact (DRIs— large single projects or subdivisions under construction or otherwise in the "pipeline")	NIPC (development database comprising projects with >10,000 square feet of nonresidential floor area *or* >10 units of residential *or* >1 acre of developed land area or development cost of >$1 million)
Grid cells (raster data layers in GIS)	ABAG (regional land use coverage— 1-hectare grid) Portland Metro (parcel data—50-foot grid) LCOG (parcel data—20-foot grid)
Sub-parcel areas (typically to capture multiple uses, occupants, or owners)	LCOG (land use "parcels") Portland Metro (vacant land areas within parcels >$\frac{1}{2}$ acre)
Sub-structural areal units	Honolulu (floors in multi-story buildings, condominiums, etc.)

How individual monitoring systems collected and represented data reflected data availability and technical resource limitations, policy needs, and the geographic scale of required analysis outputs. Within multicounty regional agencies, planners generally expressed a preference for parcel data, but questioned whether the benefits outweighed the costs.

Collection Methods and Sources

As expected, most agencies reported a strong reliance on assessors' data, especially for land records, and somewhat less for parcel GIS layers. Most agencies reported their assessors' data to be adequate, but far from ideal, for land use planning and analysis. They reported both database design and data quality problems. Overall, error rates were usually low to moderate, but uneven, with error reported especially for central city and fringe areas. Currency was an issue as well, with assessors' records lagging one to several years behind planning needs. Notably, a common view among planners was that parcel-level error, whether due to inaccuracy or to lagged currency, washed out in aggregate areawide analyses.

For systems that tracked land supply at the parcel level, use of assessors' data ranged from almost complete reliance to rejection of such data altogether. Some planning departments had undertaken independent efforts to augment assessors' land use data. For example, Seattle conducted a field survey of industrial land uses to correct and expand existing county assessor's data. Suffolk County carried out a comprehensive land use survey based on a mix of sources including assessors' records. Snohomish County updated assessor's land records with data derived from recent permits. Some agencies had abandoned the use of assessors' data altogether.

Redmond performed its entire land supply analysis with in-house inventories and permit data. The San Diego Association of Governments, which also did not use assessors' data, exemplified the limitations of assessors' data throughout California, where tax revolt legislation had limited funding for the assessment function (underscoring the connection between taxation policy and assessors' data quality).

In eight cases field survey techniques were employed to collect land use data at the parcel and subparcel levels. For example, Anchorage compiled its nonresidential land use inventory primarily by windshield reconnaissance. The Fargo-Moorhead Council of Governments used similar methods for a metropolitan-wide survey of all land uses. As a supplement to field techniques, other data (e.g., remote sensing data, and employment and land records) were commonly used to clarify ambiguous or mixed uses and to identify activities or structures that could not easily be observed from public rights-of-way.

Thirteen of the surveyed jurisdictions used image-based methods for land use change detection. These methods were employed to code individual parcels or to delineate land uses independent of parcel boundaries (using either polygon or raster cell units). For example, Portland Metro maintained its vacant lands GIS coverage through annual updates that utilized orthophotography, parcel GIS maps, and geocoded permit locations. The Metropolitan Council and the Association of Bay Area Governments each conducted regional land use inventories every five years, using primarily aerial photography. Finally, Dade County conducted a land supply analysis annually through manual inspection and classification of land use change as interpreted from aerial photography.

Eighteen land monitoring systems utilized building and subdivision permit data to augment or update other land supply information. Several of the jurisdictions surveyed had recently implemented (or were in the process of developing) automated permit tracking systems to be integrated with other components of their LIS.

Only a handful of jurisdictions reported any use of data obtained from private sources. One example was the Puget Sound Regional Council, which utilized MetroScan assessor data and information supplied by real estate brokers in order to obtain more accurate property descriptions and market data on sales activity, land availability, and prices within industrial concentrations. Another example of private-sector data use was the San Diego Association of Governments, which obtained site-level employment data from a private vendor (National Decision Systems).

A cursory review of data sharing and organizational structures revealed a heavy reliance by planning departments on data obtained from elsewhere within the same local government or from other local governments and levels of government. Within single jurisdictions, some of the newer GISs were characterized as multiparticipant "enterprise" systems that extended to all departments. Several examples of regional databases and database sharing varied in both scope and design. Among the surveyed examples of multiparticipant GIS coordination, several factors seemed to accompany successful data sharing: (1) strong regional governance structures (as evidenced by Portland Metro), (2) the establishment of formalized data sharing agreements (e.g., memoranda of understanding, interlocal agreements), and (3) a limited number of participants (such as a small number of jurisdictions within a region). On the downside, a number of planners reported significant barriers to coordination, including the independence of county assessors' departments and political fragmentation and uneven technical capabilities among local jurisdictions.

Finally, many regional data development efforts involved regional and local planning departments. Flexible, iterative protocols provided links between regional land supply databases and analyses and local information. Such cooperation functioned to correct data errors and to adjust policy and land development assumptions in accordance with local conditions. One example, the San Diego Association of Governments, distributed nine thematic "template" maps to local jurisdictions for feedback to the regional database on land use, environmental, and plan-generated geographic information (see Appendix B).

Specific Data Types

Land use data. Land use is the area in which the greatest problems with assessors' data were identified, and the area in which the most extensive in-house data capture occurred. In many cases comprehensive inventories of land use were done periodically by planning departments, but their frequency varied considerably by jurisdiction (increments ranged from one year to several decades!). Data were culled from a range of methods and sources for such inventories.

Regional land use classifications tended to be more general than local taxonomies. Generalized lookup tables commonly facilitated regional mapping of locally generated land use data, at the cost of considerable loss of potentially useful information. The San Diego Association of Governments and the Metropolitan Council were two examples of well-developed regional land use mapping programs.

Survey responses indicated the predominance of single-use classifications. However, a few efforts recorded multiple uses within parcels. For example, the Lane Council of Governments maintained a coverage of "land use parcels" that effectively split large multiuse properties, Honolulu conducted its land monitoring primarily at the subparcel level, and Kent coded parcels for primary and secondary uses as well as for land use configurations. In a small number of cases, SIC codes and state employment securities department data enhanced existing nonresidential land use inventories. Aerial photography provided data on building footprints and the spatial extent of nonstructural improvements. Occupancy data were used to track abandoned or derelict properties within central city jurisdictions that tracked land supply for economic development.

Administrative data. Zoning, administrative overlays, and planned land use were commonly maintained as separate GIS layers, which were overlaid with parcels or other data layers. A major issue for many systems was the need to rectify spatial data disparities—between parcel and administrative layers, as well as between zoning and plan boundaries and classifications.

Ownership and market data. Land ownership was commonly derived from landowner names in assessors' files. However, some planners encountered considerable difficulties in querying landowner database fields for ownership type. This prompted supplementary efforts to inventory land by ownership classification (e.g., the San Diego Association of Governments' regional land ownership survey). Valuation data were obtained from assessors' files and were considered by planners surveyed to be reliably accurate for most applications (with the one notable exception of non-tax-paying landowners). Sales data were obtained from both public and private sources. Public rights-of-way (ROW) were typically maintained as a separate GIS layer. Other restrictions of rights in land, such as easements, were obtained from land records files in both spatial and tabular forms.

Infrastructure data. A number of cases relied on parcel attribute data to describe types and levels of infrastructure servicing. Alternatively, infrastructure data were maintained as a set of GIS coverages. These included networks (line data, such as roads or sewers), individual facilities (point data, such as transit stations), and utility service districts (polygon data).

Environmental data. Environmental and physical data were typically maintained as separate GIS layers. Engineering departments commonly maintained high-resolution topographic and planimetric maps for populated areas. Federal agencies supplied layers, which included National Wetland Inventory (NWI) maps (comprising soils and hydrology data), Federal Emergency Management Agency (FEMA) maps of floodplains and floodways, Natural Resource Conservation Service soil type maps, and United States Geological Survey (USGS) maps of geomorphologic features and conditions. Occasionally, local mapping efforts improved on both the accuracy and resolution of federal data.

METHODS OF ANALYSIS

Results of the survey demonstrated that methodologies for analyzing land supply and capacity were largely ad hoc, in-house creations based loosely on best practices. Although the basic analysis steps repeated across cases, there was great variability in level of detail, variables considered, analysis assumptions, and degree to which assumptions reflected theory and professional judgment as opposed to empirical observation. Methodological designs responded to different policy requirements, development climates, types of jurisdictions, and data resources available to land supply monitoring agencies.

The following sections summarize survey findings regarding prevalent methods of analysis, discussing (1) screens used to identify and delineate buildable lands, (2) conversion of buildable lands into capacity figures, and (3) the range of development constraints considered.

Identification of Vacant, Partially Utilized, and Underutilized Lands

Buildable land supplies commonly consisted of three types of lands: vacant, partially utilized (having infill potential), and underutilized (having redevelopment potential). Lands of these three types together constituted the gross buildable land supply, prior to deductions for various constraints. (Not all agencies surveyed monitored or analyzed all three buildable land types.) Approaches to analyzing each are reviewed in turn in the following paragraphs.

Table A.6 summarizes different approaches used to screen for vacant lands.

All the jurisdictions surveyed monitored or inventoried vacant land in some form, either as the object of periodic or one-time inventories or, alternatively, as standing database elements (e.g., GIS coverages for vacant lands). Most agencies reported using primarily land use codes to identify vacant lands. Because the determination of vacant status is the most important land supply determination, one of two other measures was used to supplant or enhance assessors' data, which, reportedly, failed in many cases to capture all of the

TABLE A.6   Vacant Lands

Measures and Criteria for Identifying Vacant Land	Examples from Practice
Aerial photo observation of no improvements on land (may be carried out through GIS overlay)	Portland Metro (annual resurvey of photos identifying gross acreage and vacant areas of developed parcels >.5 acre) Dade Co. (annual parcel-level land use update and screen for buildable lands)
Parcel land use code as "vacant" or non-urban use (e.g., agriculture or forestry)—from primary or secondary data sources	Snohomish Co. (assessors' classifications) CCC (McConnell land use/cover survey maps) Anchorage (in-house land use survey)
Improvement value (none or negligible)	Seattle (improvement to land value ratio < or = .001) Kent (improvement value <$1,000)

land that planners would consider vacant. First, aerial photography interpretation was used to identify vacant land areas. Monitoring programs in Oregon (Portland Metro, Lane Council of Governments, Marion County) conducted annual aerial surveys to maintain ongoing vacant land inventories. Second, improvement value screens identified parcels classified as having a developed use (i.e., not vacant) but with only minimal or highly deteriorated structures.

Table A.7 summarizes the approaches used to screen for partially utilized lands.

Twelve of the jurisdictions indicated that they explicitly monitored infill development potential. Nearly all cases confined infill to single-family residential uses. Most of the jurisdictions employed an operational definition, amenable to database query, based on individual parcel characteristics, and without reference to surrounding development. Vacant lands inventories conducted by central city jurisdictions may also be considered infill monitoring (by definition, assessing vacant lands within a developed urban fabric), but were rarely called such.

TABLE A.7   Partially Utilized Lands

Measures and Criteria for Assessing Infill Potential	Examples from Practice
Density screens (ratio of existing to allowed density)—for single-family residential uses	Suffolk Co. (200 percent minimum lot size) Portland Metro (300–1,000 percent minimum lot size)
Percent lot coverage (ratio of building footprint or improvement area to total parcel area)—for nonresidential uses	Snohomish Co. (lot coverage <25 percent)

There were two general approaches reported in practice to identify parcels with additional land that could be developed. Density screens were most common—usually through database query for a house on a parcel that is at least twice the minimum lot size. Several agencies approached infill analysis more discriminantly by applying additional criteria to the basic screen. One factor considered was the presence of high-value single-family homes typically found on larger lots. Portland Metro, for example, screened out of its infill supply those parcels with improvement values greater than $300,000. Similarly, newly developed lots were considered by some to be unlikely to subdivide further. The Lane Council of Governments, for instance, removed from its inventory those parcels with houses built after 1970. The second general approach, percent lot cover, was applied exclusively to nonresidential uses and was not commonly adopted. For infill potential related to uses other than single-family residential, minimum lot size was used to screen out lands thought too small to accommodate additional development. For example, Snohomish County excluded parcels that were less than 1 acre in size.

Table A.8 lists measures used in practice to screen for parcels with redevelopment potential.

Seventeen jurisdictions out of 38 explicitly monitored redevelopment potential. As with infill, these analyses were carried out exclusively at the parcel level. Primary screens selected parcels with existing development that was substantially below their theoretical "highest and best use." Parcels were typically selected as characterized by either low improvement-to-land-value ratio or low ratio of existing to allowed density of development. The actual thresholds for what constituted "redevelopability" for either measure varied considerably, reflecting differences in policy, local markets, and professional judgment.

Several agencies also developed complex query methods, involving secondary screens and criteria. Montgomery County's redevelopment analysis utilized data on transit access, street frontage, utilities servicing, and impact fees, among others. Shoreline generated a "redevelopment factor" for every parcel by considering in combination both valuation and density ratios. Snohomish County, in evaluating redevelopment potential for nonresidential ("employment-based") land, considered valuation, density, and parcel size. (See Appendix B.)

As an alternative technique, land monitoring programs in Oregon employed raster-GIS-based approaches to analyzing redevelopment potential. Portland Metro and the Lane Council of Governments performed neighborhood (or "proximity") analyses to identify parcels for which the improvement-to-land-value ratio (or, alternatively, improvement value per acre) was significantly lower than the average ratio for other parcels within a radius of 500 feet. (See Appendix B.)

Redevelopment potential was also identified, based on assumptions about what types of land use conversions were likely (or not likely) to occur. For example, the Cape Cod Commission assumed that all residential uses in

TABLE A.8 Underutilized Lands

Measures and Criteria for Assessing Redevelopment Potential	Examples from Practice
Improvement to land value ratio (most common screen)—typically applied to nonresidential uses	Anchorage (high redevelopment potential—ratio of 0, moderate redevelopment potential—ratio of 0–1) King Co. (ratio <.5) Montgomery Co. (ratio <1)
Density screens (ratio of existing to allowed units or floor area)	Seattle (ratio of <0.4 for low-rise multifamily residential zones) CCC (ratio of <1—FAR standard determined by zoning, either explicitly or implicitly from setback, parking requirements, etc.) Kent (>200 percent minimum lot size—for single-family residential)
Improvement value per acre (secondary screen)	LCOG (ratio < or = $100,000)
Land use currently surface parking (secondary screen)	Newark (existing parking use as indicator of redevelopment potential) LCOG (excluded parking uses from redevelopable lands inventory)
Floor area ceiling (secondary screen)	Montgomery Co. (for commercial lands—excluded parcels that could not develop at more than their existing floor area within existing regulations)
Indicators of land assembly potential (secondary screen)	Montgomery Co. (clusters of small parcels) Boston (adjacent city-owned foreclosure properties)

commercial and industrial zones would redevelop to industrial uses. Similar use conversion assumptions determined land consumption as modeled by SANDAG's Urban Development Model. (See Appendix B.)

Planners generally regarded redevelopment to be a complex phenomenon, and methods of analyzing its potential to be relatively new and untested. Many planners described redevelopability analysis as inherently uncertain. Findings were thus frequently presented as "rough" or provisional in nature.

Approaches to Calculating Capacity

In practice, calculating capacity for residential and nonresidential land uses occurred as largely separate processes, reflecting the range of different issues influencing each type of land development. Analyses of residential capacity generally employed a linear relationship between allowed densities and build-out expectations. In the area of nonresidential analysis, the experience of surveyed jurisdictions highlighted difficulties in predicting future building utilization and land consumption rates. A less predictable relationship was reported to exist between number of employees and land utilization (or even square footage of space) than between households and building lots or dwelling units. Several agencies collected detailed demand data for nonresidential uses, such as lot coverage ratios and square feet of built space per employee by firm type (categorized by SIC code). The Metropolitan Council, the Puget Sound Regional Council, and others observed employment increases within single firms without any appreciable concomitant increase in land used (new employees were accommodated within existing facilities and sites).

Many agencies assumed that full build-out would occur on lands (both residential and nonresidential) identified as buildable, and they calculated capacity by simply subtracting existing densities from allowed densities for each use type. How planners established capacity generally reflected the purposes of the analyses. The Cape Cod Commission, for instance, estimated maximum build-out in order to gauge greatest potential environmental impacts from future development. Suffolk County, as a fast-growing suburb of New York City, considered demand for residential land to be so high within its borders as to preclude significant residential under-build.

In a number of other cases, however, agencies incorporated an explicit under-build factor or otherwise adjusted expected build-out to reflect observed development trends. For example, Fitchburg applied a 58 percent build-out assumption to all parcels planned for single-family residential development. Dade County made subarea build-out deductions based on recent land development as measured through the annual land use survey. Portland Metro applied a 21 percent residential under-build factor to all single-family parcels, and King County assessed the capacity of nonresidential uses to be 110 percent of FAR realized on recently developed sites.

Some jurisdictions also used data on observed development to break down overall demand into more specific land needs or land use proportions. For example, the Metropolitan Council employed demographic trend data to determine subarea residential demand characteristics. Portland Metro established future "refill" rates (areawide proportions of new residential and employment development occurring through infill and redevelopment). The Lane Council of Governments projected future housing type splits from existing development patterns (in this case, 40 percent single-family detached, 12 percent single-family attached, 35 percent multifamily, and 13 percent manufactured housing).

Regulatory Constraints

Few of the jurisdictions surveyed considered in their analyses of capacity any regulatory constraints above and beyond use and density limits. Seattle and the Cape Cod Commission were two cases that did; both utilized multiple zoning constraints (setbacks, height restrictions, etc.) to inform assumptions regarding determinations of effective capacity for parcels in nonresidential zones where density limits were not spelled out by regulation. Several jurisdictions reported deducting lands located within significant zoning and other regulatory overlays. For example, Seattle deducted from its inventory of buildable lands parcels in historic districts, and Portland Metro excluded dedicated farmland preservation areas.

Ownership Constraints

Most land supply inventories excluded publicly owned lands (with the notable exception of central city jurisdictions that have a large number of tax-delinquent parcels in public ownership). Also excluded were such other categories of ownership or land rights as parcels with restrictive covenants, churches and cemeteries, easements (especially utilities), and various tax-exempt uses and properties.

Physical and Environmental Constraints

Twenty-three of the cases considered environmental constraints on future development. Table A.9 lists the major types reported in the survey.

Wetlands, steep slopes, riparian zones, and flood areas were the most commonly applied constraints. However, the thoroughness of the analyses of

TABLE A.9   Physical and Environmental Constraints

Environmental and Natural Hazards Constraints on Development	Examples from Practice
Steep slopes	LCOG (>25 percent slopes) Anchorage (marginally suitable—16–35 percent, unsuitable—>35 percent)
Floodplains and floodways	LCOG (graduated constraint) Washoe Co.
Unsuitable soils	Portland Metro (flood-prone) Wake Co. (use suitability ratings)
Wetlands	Anchorage (A, B, and C classes) Metropolitan Council
Geologic and seismic hazards	Anchorage (seismic) Snohomish Co. (landslide and seismic)
Riparian corridors	Portland Metro (50- to 200-foot riparian buffers) Montgomery Co. (stream buffers)
Other water resources	CCC ("Zones of Contribution" to public well supplies)
Human-caused environmental constraints	Loudoun Co. (airport noise contours) Newark (toxic sites)

physical constraints varied considerably. Some agencies considered a single factor. Some relied on assessors' parcel descriptions to identify constrained land. Less than half of the jurisdictions surveyed conducted systematic GIS analyses of multiple physical or environmental constraints. Because of a lack of adequate data or efforts to improve data accuracy, physical factors were also addressed on a parcel by parcel basis through field surveys, rather than, or in addition to, database analysis.

Planners reported that consideration of environmental conditions was largely policy driven. In some cases, however, it also reflected unique local conditions (such as avalanche hazards in Anchorage) or pressing concerns about environmental impacts (such as on groundwater in Cape Cod). Environmentally noxious conditions caused by past or adjacent land uses also constituted land development constraints.

Environmental factors functioned most often as outright barriers to development—or as prohibitive constraints. In several cases, planners applied the concept of a partial or mitigating constraint, identifying lands where development may occur, but at a lower than expected density. The Lane Council of Governments, for example, discounted the effective capacity of parcels that fell within floodplains (but not floodways), some classes of wetlands, hydric soil areas, and on slopes greater than 25 percent. Other jurisdictions established suitability ratings based on multiple factors, such as Anchorage, which assigned one of three suitability ratings to parcels based on topography, hydrology, and natural hazards.

In lieu of site-level information on environmental conditions, some cases applied generalized deductions to aggregate land supply figures. For example, lacking adequate GIS resources to conduct thorough assessments, some jurisdictions in Washington applied blanket percentage deductions for critical areas within subareas or even entire jurisdictions. Overall, environmental constraints were applied differently depending on land and development context—with vacant lands often discounted more heavily than infill and redevelopment sites. Several central city jurisdictions surveyed (e.g., Seattle) did not consider environmental constraints at all.

Notably, environmental factors were also evaluated from the perspective of carrying capacity, whereby future development impacts on critical resources and systems were measured against politically or scientifically determined thresholds. For example, the Cape Cod Commission combined development build-out with areawide environmental impact analyses.

Infrastructure Constraints

Fifteen cases out of 38 considered infrastructure factors in their land supply analyses. Table A.10 describes the three major categories—public use deductions, screens for serviced land, and carrying capacity measures—with illustrations from practice.

TABLE A.10   Infrastructure Constraints

Approaches to Dealing with Infrastructure Constraints on Development	Examples from Practice
Area deductions for future needs	Street rights-of-way: PSRC (deduction of 10 percent for vacant, 5 percent for partially developed lands) Public facilities: Snohomish Co. (15 percent deduction); Redmond (0–5 percent) Open space/parks: Boulder Co. (unspecified); Portland Metro (projected service need deductions for streets, schools, parks, churches, etc.)
Screening for serviced lands	Utilities: Wake Co. (serviced lands identified through utilities buffers); CCC (coded industrial areas for full range of utilities, including fiber-optic) Transportation: Newark (parcel by parcel analysis of access to transit and highways); Anchorage (accessibility rating through graduated 100- to 500-foot buffer to roads by class)
Carrying capacity considerations	Adequate Public Facilities Ordinance (APFO): Montgomery Co. (transportation and school enrollment capacity within service areas) Regional infrastructure capacity: CCC (compared buildout with transportation system capacity)

Deductions often depended on development context (greenfields development were discounted more heavily than infill and redevelopment) and parcel size (large parcels were discounted more than small parcels, particularly for ROW). GIS played an important role in relating land data to utilities infrastructure, with reported examples of network buffers to establish parcel serviceability as well as service area polygons overlaid with parcels to relate potential land development to levels of service capacity.

Applications of a Market Factor

Several examples of the use of "market factors" were reported (all in Washington, where state and county guidelines recommended the practice). The magnitude of the market discount varied by jurisdiction and land supply category. For example, Kent deducted 5 percent from vacant lands capacity and 15 percent from redevelopment capacity; and King County—5 to 15 percent for vacant lands, 10 to 15 percent for redevelopable lands. Alternatively, Seattle deducted 15 percent from the capacity yielded by low-rise redevelopment, single-family infill, and vacant land development, with no market factor applied to the remaining buildable parcels that were identified using valuation criteria (assumed to adequately capture market forces). Snohomish County deducted 15 percent from vacant lands capacity and up to 50 percent from employment-based land use capacity. Notably, land monitoring agencies in Oregon had previously employed market factors of approximately 25 percent. However, planners reported that such practice was ruled invalid by state courts in favor of empirically based land market assumptions.

Appendix B

Selected Case Summaries

The following are summaries of eleven cases derived from the national review of land supply monitoring programs. The first three summaries—Portland Metro, Oregon; State of Maryland; and Montgomery County, Maryland—overlap with and supplement Chapters 3 and 4. The subsequent eight summaries—Buildable Lands Program, State of Washington; City of Seattle, Washington; Snohomish County, Washington; Metropolitan Council, Minneapolis–St. Paul Minnesota Cape Cod Commission, Massachusetts; Municipality of Anchorage, Alaska; San Diego Association of Governments, California; and Lane Council of Governments, Oregon—provide detailed information about other programs that represent a range of approaches to land supply and capacity monitoring, encompassing both parcel- and non-parcel-based geographic information systems (GIS) applications. Generally, the summaries address, as separate elements, regional and policy contexts, GIS resources, and, most prominently, methodologies for monitoring and analyzing land supply and capacity.

Summaries are by Michael Hubner, with the exception of the Summary of the Buildable Lands Program, Washington, which is by Shane Hope, Managing Director, Growth Management Services, Washington State Department of Community, Trade, and Economic Development.

Monitoring Land Supply with Geographic Information Systems, edited by Anne Vernez Moudon and Michael Hubner ISBN 0 471371673 © 2000 John Wiley & Sons, Inc.

CASE SUMMARY: PORTLAND METRO, OREGON

Overview

Portland Metro is the regional planning organization for the Portland metropolitan region. Its jurisdictional area includes three counties (Multnomah, Clackamas, and Washington). The regional population in 1994 stood at more than 1.5 million. As part of a long-standing program of state land use planning in Oregon, Metro has maintained an Urban Growth Boundary (UGB) as the primary tool of growth management for the past 20 years. At present, the UGB (for the three Oregon counties) contains approximately 232,000 acres (362 square miles) of land in various states and densities of development.

Several factors have fueled the efforts of Metro to develop a comprehensive regional system for monitoring land supply. The UGB strategy has generated concern about the potential effects of land supply constraints on land and housing prices. Recently, a sustained period of economic and population growth, coupled with lower than expected development densities within the UGB, have spurred multidimensional approaches to measuring the capacity of land throughout the region.

State of Oregon statutes require that every five years Metro review and update its estimates for the capacity of the land within the UGB to accommodate the next 20 years of anticipated growth. Metro's most recent such effort, the "Buildable Lands Inventory and Capacity Analysis," was completed in 1997 and published as part of the *Urban Growth Report*. The findings of this analysis provide information for the implementation of the *Regional Framework Plan*. The plan includes the Region 2040 Growth Concept, which seeks to increase the capacity of land inside the UGB through zoning changes, transit planning, and incentives for higher-density development within designated centers and corridors. These efforts are taking place amid criticism regarding the accuracy of previous capacity analyses done by the agency, fueled by a concern that the UGB is unduly constraining the market for land and leading to housing price inflation.

Geographic Information System Resources

The Data Resource Center, located within Metro's Growth Management Services Department, maintains the Regional Land Information System (RLIS), a comprehensive, multipurpose GIS. RLIS contains more than 100 data layers, including parcels, zoning, comprehensive plan areas, parks and open space, soils, wetlands, topography, land cover (vegetation), and floodplains (100-year).

Local jurisdictions provide data for Metro to update its parcel base map. The parcel layer is linked to attribute data drawn primarily from county assessors' databases. The parcel boundaries are updated on a continuous basis. Metro staff describe the quality of the parcel data as good, although they do report some problems related to the timeliness of base map revisions, especially for the fastest-growing areas in the region. Metro has established a common set of zoning and plan designations, which serves to integrate within the regional database all of the various zoning and land use coding schemes used by local jurisdictions. The Data Resource Center archives land records data via time-stamping database additions and updates as they occur. A "snapshot" of current conditions is generated annually and archived for future reference.

The Data Resource Center also maintains a coverage of undeveloped lands, which is updated annually. With the previous year's undeveloped lands layer used as a base, revisions are made using information from digital orthophotography (visually checked against a plot of the old map layer), updates of tax-lot boundaries, and building permits for the previous 18 months to flag areas of likely change.

Land Supply Monitoring and Capacity Analysis

Buildable Lands Inventory and Capacity Analysis (1997)   As described in the *Urban Growth Report* (Portland Metro 1997), Metro developed a 14-step methodology to estimate capacity of the land supply within the UGB to accommodate both residential and employment uses— the "Buildable Lands Inventory and Capacity Analyses." The following is a summary of the steps in that analysis. Steps 1 through 8 are referred to as the "standard methodology" that had been followed previously; Steps 9 through 16 are additional, addressing new considerations and targeted to support the implementation of the new 2040 Growth Concept.

Step 1: Calculate the total number of acres inside of the Metro UGB.

Step 2: Subtract acres of developed and committed lands to arrive at total gross vacant acres.

Step 3: Subtract acres of platted, vacant single-family residential land.

Step 4: Subtract vacant, environmentally constrained acres to arrive at gross buildable vacant acres. Constraints include the following:

- Slopes >25 percent
- 100-year floodplains
- Flood-prone soils
- Wetlands
- 50- to 200-foot riparian buffers

Step 5: Subtract land for future public facilities to arrive at net buildable vacant acres (referred to as "gross-to-net reduction"). Included in this deduction are the following:

- Streets deductions, dependent on parcel size—22 percent for parcels >1 acre, 10 percent for parcels <1 acre

- Schools deductions based on expected population increase within UGB
- Parks deductions based on the ratio of parklands to expected future population increases

Step 6: Calculate dwelling unit and employment capacity of net buildable vacant acres under current comprehensive plans (and associated zoning). Assumes build-out to allowable densities.

Step 7: Deduct housing units from current comprehensive plan capacity to reflect single-family under-build. Lands planned for single-family uses are thus expected to build out at 21 percent below maximum allowable densities.

Step 8: Adjust dwelling unit and employment capacity (down) to reflect undeveloped platted lots and for development rights on unbuildable land. (Adds back what was subtracted in Step 3.)

Step 9: Rezone buildable parcels (from Step 4) to reflect 2040 Growth Concept and recalculate dwelling unit and employment capacity. (This operation is performed with the use of a rasterized parcel data layer.)

Step 10: Discount household and employment capacity under the Metro 2040 Growth Concept to reflect residential and employment under-build. Build-out is projected to be 21 percent for most residential areas and variable for employment uses, determined on a parcel-by-parcel analysis of development constraints such as small parcel size, restricted access, steep slopes, or existing partial development.

Step 11: Adjust planned density assumptions to account for the time needed for cities and counties to fully implement zoning changes required by the Urban Growth Management Functional Plan (referred to as "ramp-up").

Step 12: Estimate redevelopment potential and adjust capacity calculations for both dwelling units and employment. This is accomplished by using an improvement-to-land-value ratio screen for larger parcels. For smaller parcels, a comparison of improvement-to-land-value ratio with those of surrounding parcels is done by utilizing a neighborhood function with rasterized parcel data.

Step 13: Estimate infill housing and employment absorption and adjust capacity accordingly (based on an estimated rate of infill within a stock of oversized lots—at least 300 percent minimum size allowed under zoning).

As reported by Metro planners, neighborhood-level monitoring of construction activity has corroborated the estimates of infill capacity derived from the methodology, including identified infill development potential in some of the higher-income neighborhoods, areas that had been thought to be stable and essentially built-out.

Step 14: Adjust dwelling unit and employment capacity to reflect existing platted lots and development rights on unbuildable land (same as Step 8).

Notably, under the new methodology (Steps 9 through 14) the conversion of parcels from vector to raster (approximately 50-foot grid) entails a small decrease in site-level accuracy to estimate development potential. Some of the analyses made after the conversion are more approximate than they might have been if performed directly with the parcel-layer polygon. However, working with raster data reduces the computational burden of upzoning lands regionwide, especially given the inconsistencies in parcel attributes assigned by local jurisdictions. Moreover, although the regional analysis of buildable lands capacity converts some of the data into raster format, other smaller area studies have employed a parcel-based approach to capacity analysis.

An explicit market factor is not employed by the Metro capacity analysis. Market influences are implied by the adjustments made for under-build. Otherwise, market considerations are embedded within the demand forecasts made by Metro.

Baseline Urban Growth Data (1997)   The *Baseline Urban Growth* Data report provides baseline data for a monitoring program under development to track growth and land use change. Some of the data analyses presented in the report may eventually be used by Metro to generate Performance Measures of the achievement of various regional planning objectives. Several measures presented in the report relate to aspects of land supply, as follows:

- *Vacant land conversion.* Measures of the rate of consumption of vacant land and the distribution of land uses to which it is converted. Data sources: Metro Vacant Lands Inventory, RLIS, population and housing data from several sources.
- *Housing development, density, rate, and price.* Median lot size for new single-family houses within the UGB and units per acre for multifamily housing construction. Price measured as average sales amount for single-family, average rent for multifamily. Data sources: Building permit records, assessors' sales ratio studies, rental surveys, RLIS.
- *Infill and redevelopment.* Rate of infill as a percentage of total new housing and employment growth (aggregated by land use). Data sources: Building permit data, RLIS, *Real Estate Report for Metropolitan Portland, Oregon* (Portland Metro 1996).
- *Environmentally sensitive lands.* Assessment of the amount of such land that is permanently protected and the amount that is developed. Data sources: Wetlands inventories, slopes data, floodplain data, soil surveys, aerial photography, RLIS.
- *Price of land.* Cost of land based on lot sale price (limited to sales of vacant land). Data sources: Assessors' records, RLIS.

The report also contains detailed information about initial analyses performed to establish the baseline figures for these categories. Parcels were a crucial input in all of the analyses—either as the units for measuring land use change or status, or as the units for measuring valuation.

Contacts

David Ausherman
Glen Bollen
Jennifer Bradford
Sonny Condor
Portland Metro, Data Resource Center

Sources

Portland Metro. *Baseline urban growth data—Preliminary review draft.* April 1997.

———. *Urban growth report—Final draft.* December 1997.

———. Metro web site available at [http://www.multnomah.lib.or.us/metro].

CASE SUMMARY: STATE OF MARYLAND

Overview

Maryland is one of the most urbanized states in the country. Its relatively small size (12,297 square miles) and high overall population density (512 persons per square mile) have helped focus state-level attention on issues related to land use and land supply in urban and urbanizing areas. At the same time, significant parts of the state are still devoted to productive agriculture, a traditional land use that is at risk of displacement by rapidly spreading urban development. At present, more than 85 percent of the state's five million residents live in the urban corridor between and including Baltimore and the Washington, D.C., suburbs.

Reflecting Maryland's small land area, planning and land management functions have a more prominent place in state government here than in many other states. Property assessment is a state-level function, and, notably, the Department of Assessments and Taxation archives property records and parcel base maps for the entire state. The Maryland Office of Planning has taken a strong role in offering planning assistance to local governments, as well as in conducting statewide planning studies related to land use and growth capacity.

State growth policy emphasizes urban containment for resource preservation and fiscal savings related to infrastructure. It also addresses issues related to the economic decline of older cities. The latest efforts by the State of Maryland to address the state's growth issues represent an extension of work by a few local jurisdictions to limit urban sprawl.

State Guidelines for Smart Growth Compliance/Priority Funding Areas

The Office of Planning is currently engaged in implementation of the Economic Growth, Resource Protection, and Planning Act of 1992. The goals of the Act include fostering concentrated development, protecting sensitive areas, and preserving rural resource areas. Part of the Office's mission is to provide technical support to local jurisdictions, particularly as related to growth management implementation and environmental protection. Further, the Smart Growth Areas Act of 1997 builds on the earlier legislation by providing financial and other incentives to local governments to limit sprawl and preserve rural lands. A complementary goal is the revitalization of declining urban areas by directing increased funding to development within existing communities.

In compliance with state funding provisions for infrastructure and other improvements, local jurisdictions must designate Smart Growth Areas. Such areas may include the following:

- Existing municipalities
- Areas inside the major metropolitan beltways
- Existing enterprise zones, neighborhood revitalization zones, heritage areas, and industrial lands
- Areas designated by local jurisdictions as "Priority Funding Areas." (PFAs)

The Act gives local governments flexibility in delineating PFAs. The Office of Planning has issued guidelines for jurisdictions to make this delineation in compliance with state law. *Smart Growth: Designating Priority Funding Areas* (1997) presents a set of "definitions and methodologies" for "calculating land capacity, future land needs, and residential densities" within designated PFAs. In addition to providing guidance for conducting a land capacity analysis (partially based on practices in Oregon and Washington), this document also gives technical guidance on simple population forecasting, measurement of residential density (a key criterion for PFAs), and the designation of rural villages as PFAs. The focus of state oversight is not to establish a set of strict guidelines but to encourage consistency between local jurisdictions through technical assistance (especially to jurisdictions that lack resources or experience in this area).

MdProperty View Database

MdProperty View is a statewide parcel-level GIS product made available on CD-ROM to all of the counties in the state for their use in complying with the Smart Growth legislation. The database can be viewed and manipulated with ArcView software. MdProperty View is targeted for use by counties in designating Priority Funding Areas and in performing other work related to Smart Growth.

The centralization of property assessment within one agency in Maryland made the development of a statewide parcel GIS database possible and relatively easy to produce. It allowed Maryland to avoid the land monitoring database problems often encountered in combining local assessors' files: jurisdictional inconsistencies in parcel attributes (especially land use) and the uneven timeliness of data updates resulting from staggered batch processing of land records. The Maryland State Department of Assessments and Taxation provides the counties with complete parcel attribute files from its Computer System Mass Appraisal database. Consistency in parcel identifiers is thus enforced by the state for all jurisdictions, regardless of whether they are using MdProperty View.

MdProperty View represents the parcels as a point coverage (parcel centroids) with an overlay of parcel boundaries derived from scanned assessor's maps. Some counties, such as Montgomery County, had previously created vectorized parcel layers and other coverages for their GISs. However, for counties without prior GIS experience or capabilities, the new state product greatly enhances the ability to assess land supply and capacity, while allowing for flexible small-scale aggregation of the data. For some areas, layers in addition to parcels are also available from the state, including data on floodplains, wetlands, agricultural easements, watershed boundaries, digital elevation models, critical areas, land use/land cover, and other factors.

Statewide Capacity Analysis

The State of Maryland performed its own statewide capacity analysis in the early 1990s—summarized in "The Potential for New Residential Development in Maryland: An Analysis of Residential Zoning Patterns" (Maryland Office of Planning). The purpose of this study was to "examine how Maryland's counties are using zoning and sewerage to meet the demands for growth." The analysis was done using the state's GIS, which at the time included land use, zoning, and sewer service area coverages, but did not include a parcel layer. The Office of Planning reports that this earlier capacity analysis would have been more easily performed at the parcel level. If and when the state revisits large-scale capacity analysis, it will most likely employ the database contained in MdProperty View.

Contacts

Mike Leitre
Jim Noonan
Maryland Office of Planning

Sources

Maryland Office of Planning. 1997 *MdProperty View—Program description and documentation.*

———. 1997b. *Smart Growth: Designating priority funding areas* 1997. At [http://www.op.state.md.us/].

———. 1998. "The potential for new residential development in Maryland: An analysis of residential zoning patterns." (m.d.) *Managing Maryland's growth issue papers.* At [http://209.116.10/mmg/mmg.htm].

CASE SUMMARY: MONTGOMERY COUNTY, MARYLAND

Overview

Montgomery County, one of the "collar" counties surrounding Washington, D.C., has experienced high growth pressures for a number of years. From a population of 805,930 in 1995, the county is expected to reach a total of over 1,000,000 residents by the year 2020 (Montgomery County 1997). The county's 492 square miles encompass suburban communities, in both an inner and an outer ring, as well as rural areas. Before the recent development of a parcel-based GIS (PBGIS), Montgomery County had a long track record of monitoring land supply using mapped infrastructure service areas. Recent work focusing on land supply for economic development has utilized parcel-level data (see the description of nonresidential capacity analysis under "Site Characteristics Inventory"). The county now plans to integrate PBGIS with its broader growth management activities.

Montgomery County's slow recovery from the recession of the early 1990s has been a driving force behind recent planning efforts. The county lags behind some of its neighbors in new construction and land redevelopment. The latest "Annual Update" on the local economy indicates that the county has had "less construction activity than the low vacancy rates, rising rents and number of approved pipeline projects would suggest." Private-sector critics of local growth management have pointed to land supply constraints as a factor hindering economic recovery and development. The private development sector has applied pressure to have restrictions relaxed on approximately one third of the land in the county that is precluded from development because of agricultural set-asides.

Montgomery County's land information system (LIS) is one of the primary cases studied by Godschalk et al. (1986). The planning department's main use of the system at that time was for monitoring land supply related to the county's Adequate Public Facilities Ordinance (APFO). This effort utilized assessor's parcel records and pipeline development files aggregated to "Policy Areas". To administer the APFO, the county compared existing development and that in the pipeline development (for each Policy Area) with a threshold level established in coordination with capital facilities planning. The output of this analysis helped decision-makers and planners to impose development restrictions and regulations that were responsive to supply and demand within each sub-area. Many of the features of this system still exist.

Geographic Information System Resources

MC:MAPS is an inter-agency entity that manages a comprehensive multi-participant GIS for Montgomery County. A parcel coverage for the entire county was completed in the summer of 1997. This parcel coverage is at a higher resolution than that provided by the state in the MdProperty View and offers the advantages of working with vectorized parcels.

Other GIS layers maintained by MC:MAPS include orthophotography, topography, buildings, planimetric features, soils, floodplains, and utility districts. For the purposes of tracking pipeline development, MC:MAPS maintains a separate GIS layer for permits and subdivisions. This data layer is spatially compatible with the parcel layer for the purpose of overlay analysis. The attribute data for the parcel coverage include land and improvement valuation, zoning, land use classification, building height in stories, utilities servicing, tax category, and status of current permits (forthcoming). Extracts of state assessor's files are used as the primary source for parcel attributes.

Most of the recent and ongoing GIS efforts by the county have focused on database development, especially in generating the various environmental layers and a master address file system. In addition, the county has been compiling historical data (records from 1985 up to the present) on initial development applications, approved site plans, records of government projects, and water and sewer infrastructure improvements or extensions. County planners hope to be able to use this legacy database for trend analyses related to future land development.

Land Supply Monitoring and Capacity Analysis

Enforcement of the Adequate Public Facilities Ordinance (APFO) within designated Annual Growth Policy (AGP) areas is a long-standing application of Montgomery County's land information system. An annual report updates the status of each AGP area. Transportation-based ceilings are established for all AGP areas, with gross capacity defined as "the total amount of development (existing, approved for construction, yet-to-be-approved) that can be handled by the transportation network" (Montgomery County 1997a). The analysis is also sensitive to the potential effect of jobs/housing mix on transportation demand. Total numbers of households and employees (both existing and in the pipeline) are subtracted from the gross ceiling for each area to determine net remaining infrastructure capacity.

Although transportation capacity functions as the primary supply constraint, the service-level capacities of schools and of other utilities infrastructure are also taken into account. For some AGP areas, the staging ceilings are implicitly set at the prevailing zoned development capacity. Where the level of existing and pipeline development exceeds the gross ceiling, the county has imposed development moratoria. (The county is in the process of relaxing APFO rules to allow for a pay-as-you-go policy for many areas. A planned four-year hiatus for the program is intended to stimulate new development activity.)

The county's last residential land capacity study was done in 1992, prior to the implementation of a parcel-based GIS. At the urging of the local builders association, the planning department expects to use the new GIS data layers to update the housing database for the entire county. This will include an inventory of both vacant and redevelopable lands that can accommodate residential uses.

Site Characteristics Inventory (1998) Meanwhile, in response to concerns about the local economy, the first application of the new parcel-based system has focused on industrial, office, and commercial land use capacity—summarized in the *Site Characteristics Inventory Report.* In this analysis, planners developed suitability indices corresponding to various physical and environmental constraints to screen parcels for classification as either vacant or redevelopable. A private consulting firm, EDAW Integrated Solutions Group, provided technical support for the detailed analysis of site constraints. (According to county planners, the Planning Commission decided to contract with an outside firm to enhance limited in-house resources and to satisfy private-sector critics of the accuracy of initial in-house analyses.) To estimate future development capacity, the resulting inventory was matched with corresponding plan designations that indicated the type and intensity of uses allowed on each parcel.

The *Site Characteristics Inventory* entailed a secondary and detailed analysis of the sites identified in an earlier phase as having development potential. The study consisted of screening for environmentally nonbuildable areas (wetlands, steep slopes, stream buffers), as well as for commercial projects currently in the pipeline. The intent of the study was to "provide a more definitive snapshot of the quantity of approved and developable properties that could facilitate economic development and master plan build-out." Data sources included files on projects in the development pipeline, a countywide vacant lands inventory, and an inventory of potentially redevelopable lands in central business districts. The analysis identified vacant and redevelopable parcels for each "community-based area" and incorporated data on zoning, parking, traffic management districts, impact fee areas, proximity to transit stations, road frontage, and water and sewer servicing.

The *Inventory* is of special interest in the following areas:

- To identify redevelopable parcels, the analysis employed an improvement-to-land-value ratio of 1.0. The redevelopment analysis excluded parcels that could not develop at more than their current floor area. An outside real estate consultant also reviewed the selected parcels to further screen out properties that faced additional barriers to economically feasible development.

- For partially constrained parcels, the analysis took into account the potential for transferring development density allowances from the unbuildable to the buildable portion of the site.

- The study found a large number of small parcels with redevelopment potential and attempted to assess the potential for land assembly as a determinant of future development capacity. Interestingly, the construction of a new subway has spurred intense interest in redevelopment possibilities around station areas. Although redevelopment in some of the older central business districts (CBD) has been hampered by small parcel sizes, it is hoped that the results of the analysis will help facilitate land assembly and revitalization.

- The developability criteria applied by EDAW included both environmental factors and the economic feasibility of new development. Using the GIS parcel and administrative boundaries, and environmental data layers provided by the county, EDAW identified and delineated environmental constraints that included slopes, soils, and critical drainage areas. The results provided detailed information about buildable vacant and redevelopable lands, accounting for environmental constraints, specific growth management programs, impact taxes, and other factors affecting development potential.

Given concerns about the economic stagnation of "inner-ring" suburban areas, the study was intended as a step toward explaining the disparity in economic development between Montgomery county and its adjacent counties, and a lead to suggesting economic development strategies. The next phase of the capacity analysis will explore how this inventory relates to the needs of employers as potential users of nonresidential lands. The end result will be the crafting of a set of "strategic recommendations on any action indicated to better coordinate and conform site availability with economic objectives" (Montgomery County Planing Board and MNPPC 1998b, 2).

Additional Land-Monitoring-Related Activities Using Parcel-Based GIS

Montgomery County has begun using its parcel-based GIS for analysis in support of a transfer of development rights (TDR) program, to screen the parcels of both potential senders and receivers of development rights for possible inclusion in the program based on suitability criteria. The county used the parcel data for assessing the current level of compliance with the state's new Smart Growth policy. This work entailed verifying the match between state-designated and county-recognized urban areas and communities (areas of high and low population density and delineated "growth areas"), as well as the designation by the county of Priority Funding Areas conforming to state-defined criteria.

Parcel-level capacity analyses were also recently done in support of sub-county comprehensive plans for the Silver Springs and Potomac jurisdictions.

Contacts

Andrew Perez, Research Planner

Sally Roman, Research Coordinator

John Schlee, Planner

Apollo Teng, GIS Manager

Montgomery County Department of Park and Planning, Research and Technology Center

Brad Dailey

EDAW, Integrated Solutions Group

Sources

Godschalk, David R., Scott A., Bollens, John S. Hekman, and Mike E. Miles, 1986. *Land supply monitoring: A guide for improving public and private urban development decisions.* Boston: Oelgeschlager, Gunn & Hain, in association with the Lincoln Institute of Land Policy.

Metropolitan Washington Council of Governments web site at [http://www.mwcog.org].

Montgomery County. 1997. Official web site. Available 19 October 1997 at [http://www.clark.net/pub/mncppc/montgom].

Montgomery County Planning Board and the Maryland-National Capital Park and Planning Commission. 1995. *Montgomery County's Annual Growth Policy.* November.

———. 1997. *Annual Growth Policy ceiling element—Fiscal year 1999.* November.

Research and Technology Center. 1998a. *Economic forces that shape Montgomery County—Annual update.* February.

———. 1998b. *Site characteristics inventory.* Montgomery County Department of Park and Planning. February.

"BUILDABLE LANDS" AMENDMENT TO WASHINGTON STATE'S GROWTH MANAGEMENT ACT: PROGRAM SUMMARY

Shane Hope

A program to monitor land supply and growth management objectives is underway in Washington State. Although still in its early stages, the "buildable lands" program is extensive and dynamic. This chapter examines why it was started and what key issues are emerging.

Background

In Washington State, growth management legislation was passed in 1990. This legislation, known as the Growth Management Act (GMA), establishes general goals and specific requirements for future land use. Although the full set of GMA requirements currently applies to only two-thirds of the State's cities and counties, these jurisdictions contain approximately 95 percent of Washington's population. A minimum set of requirements applies to all other cities and counties.

Under GMA, 13 goals are established, encompassing urban growth containment, sprawl reduction, economic development, environmental protection, provision of adequate public infrastructure and affordable housing, as well as other statewide priorities. Implementation of the goals occurs primarily at the local government level, through a framework that includes 1) countywide planning policies, 2) comprehensive plans, 3) development regulations, and 4) capital budgets and other ongoing activities. Urban Growth Areas must be adopted to include existing incorporated urban areas, as well as appropriate locations for future urban expansion. Lands outside of the Urban Growth Boundaries (UGBs) are to be devoted primarily to rural or natural resource activities, including agriculture, forestry, and mineral extraction. The mandatory designation of critical areas will encompass environmentally sensitive lands.

GMA also requires that local governments conduct a through review of local plans and regulations every five years. Under a 1997 amendment (HB6094), six of the most populated counties and the cities within them (currently a total of 102 jurisdictions) also must comply with special requirements for monitoring land supply and urban densities. The requirements, known collectively as the "Buildable Lands Program" were based, in part, on the concern of developers that UGBs might overly limit future development opportunities and, in part, on the concern of environmentalists that resource lands and critical areas might not be adequately preserved. Both sets of concerns implied a need for accurate information to assess the effectiveness of local governments in implementing GMA. The Buildable Lands program lays out the process for collecting and analyzing that information. State oversight will emphasize coordination and consistency in data collection and methodologies, especially between the counties and their constituent cities.

Program Requirements

The Buildable Lands requirements currently apply to six counties: Clark, Thurston, Kitsap, Pierce, King, and Snohomish Counties, encompassing more than 7,600 square miles. The program's primary purposes are to:

1. Evaluate whether local growth targets and objectives are being met, especially for achieving urban densities within UGBs
2. Ensure actions are taken, other than adjusting UGBs, to provide for an adequate supply of land, while still meeting other requirements of the GMA

Local governments must take the following steps:

1. Adopt a comprehensive plan that incorporates baseline information about expected population growth and future residential- and employment-based land use needs for the next 20 years.
2. Collect development data annually, using consistent procedures and methods (as appropriate and feasible).
3. Evaluate the data at five-year intervals, starting in 2002.
4. At each five-year interval, identify and implement actions that should be taken to increase consistency between development trends and local plans.
5. Determine whether the actions taken to increase consistency have been effective and adjust the action plan as necessary.

An appointed state agency, the Department of Community, Trade, and Economic Development (CTED), provides technical assistance to guide the process, makes grants to affected cities and counties, and prepares periodic progress reports for the Legislature.

Questions Addressed by the Program

HB6094 is specific about requiring that certain questions be answered through the local jurisdiction land monitoring process. These include:

- Is there sufficient suitable land to accommodate expected 20 years growth?
- What quantity and type of land is needed to meet future demand?
- Are urban densities being achieved inside UGBs?
- What are the densities of housing that has been constructed during the evaluation period?
- Within the UGBs, how much land was developed for commercial and industrial uses during the evaluation period?
- Based on the observed development levels, how much land will be needed to meet expected demand over the remaining portion of the 20-year planning period?

Other questions, while not spelled out precisely, are strongly implied by the statute. These include:

- Are there inconsistencies between planned and actual land use patterns? If so, what are they? Why are they occurring?
- To what extent are capital facilities available—or needed—for land to be developed?
- To what extent do environmental constraints affect future land development? What actions have been taken by the local jurisdiction to meet land development objectives?

Program's Principles

A set of principles guides the implementation of the Buildable Lands program. These include:

- Data collection and evaluation should build on the work that local governments have already done.
- Common definitions and compatible data should be used whenever possible.
- Partnerships between and coordination among jurisdictions, especially counties and their constituent cities, are essential.
- New guidance from CTED will expand on and refine previous information (e.g., the GMA guidebook "Providing Adequate Urban Area Land Supply") that was issued to assist local governments in the initial UGA design process.
- Choices made about methodologies for monitoring and evaluating of land supply must be tempered by the realities of funding.
- The program will evolve over time, as more information and resources become available.

Current Status of Program

An initial survey of jurisdictions by CTED revealed many inconsistencies and gaps in the land supply data that are currently collected, as well as uncoordinated efforts in developing land information systems among the various agencies involved. Since then, CTED representatives have been meeting with an advisory committee (comprised of representatives from regional and local agencies) to develop recommendations for monitoring land supply and urban densities under the GMA. Differences in expectations among the 102 jurisdictions involved have presented challenges for the committee. Funding limitations have also added difficulties in the implementation of the Buildable Lands program. However, a preliminary set of definitions and guidelines is now complete. Counties and cities are already working to refine their GIS data and to gather additional data on development activity.

Working Definitions

Working definitions used in the guidelines include the following.

Growth Target: A figure in an adopted policy statement indicating the type and amount of growth (e.g., number of persons, households, or jobs) a jurisdiction intends to accommodate during the planning period.

Key Development Data: Information that is critical to identify the location, timing, and scope of new development that has occurred. Components may include, but are not limited to, building permits, certificates or changes of occupancy, subdivision plats, zone changes, urban growth boundary amendments, numbers of dwelling units, and critical areas and related buffers.

Lands suitable for development: All vacant, partially used, and under-utilized parcels that are (1) designated for commercial, industrial, or residential use; (2) not intended for public use; (3) not constrained by critical areas in a way that limits development potential and makes new construction on a parcel unfeasible.

Partially used land: Partially used parcels are those occupied by a use but contain enough land to be further subdivided without need of rezoning. For instance, a 10-acre parcel on which there is a single house, where urban densities are allowed, is partially developed.

Sufficient land supply: Amount of land necessary to accommodate adopted population and employment forecasts or targets for the 20-year planning period, taking into account any appropriate factors.[1]

Underutilized land: All parcels of land zoned for a more intensive use than that which currently occupies the property. For instance, a single-family home on multifamily-zoned land is generally considered underutilized. This classification also includes redevelopable land—that is, land on which development has already occurred but on which, because of present or expected market forces, there exists the strong likelihood that existing development will be converted to more intensive uses during the planning period.

Vacant parcels: Parcels of land that have no structures or that have buildings with very little value.

General Approach to Data Collection

Different technological approaches have been considered for monitoring the buildable land supply. A computerized geographic information system (GIS) offers many elements that make it the most practical approach to integrating

---

[1] See Susan C. Enger, *Issues in Designating Urban Growth Areas,* part 1, *Providing Adequate urban Area Land Supply* (Olympia, Wash.: State of Washington Department of Community Development, 1992).

local data and meeting the statutory requirements. GIS allows land use information and land ownership information to be combined and viewed graphically. All six of the counties covered by the legislation, and a few of the cities within them, already have GIS. Aerial photography is also useful to collect data for the assessment of land supply, and this method may supplement GIS. However, the costs associated with high-resolution remote sensing imagery are currently too prohibitive for it to be used regionwide. Although county assessors' data are neither completely accurate nor up-to-date, they offer the best available starting point for building parcel databases to monitor and analyze land supply.

Eventually, either the counties or regional planning councils will maintain countywide GIS coverages on land and land development (with much of these data compiled from county and city sources). To this end, local governments currently collect data on building permits, critical areas, and other geographic information. Agreements are being developed between counties and cities to establish coordinated data collection and storage efforts. Data collection that goes beyond that necessary to meet legal requirements may be performed by individual jurisdictions or through a coordinated intergovernmental process.

Measures to Achieve Growth Objectives

Part of CTED's role has been to issue a list of suggested measures or techniques for local governments to encourage higher residential densities and employment-based development within UGBs. This list includes the consideration/implementation of the following:

- Accessory dwelling units (e.g., "granny flats")
- Capital facilities investments in urban areas
- Clustering development
- Cohousing
- Density bonuses
- Downtown revitalization
- Duplexes and town homes
- Economic development strategies
- Efficient transportation (with emphasis on walking, biking, and transit)
- Environmental review and mitigation performed by communities that want new urban growth to locate in special planning areas
- Industrial reserve zones (to be held for future industrial uses)
- Maximum lot sizes
- Minimum density requirements
- Mixed uses
- Multifamily housing tax credits

- Phasing urban growth to match infrastructure
- Small-lot zoning (e.g., 2,500- to 6,000-square-foot lots for detached housing)
- Transfer of development credits
- Urban amenities that support increased densities
- Urban centers and villages

The list can also be used as a guide for local governments in meeting the fourth step of the Buildable Lands requirements—taking corrective actions.

Major Questions Still Being Considered

Many key questions have been considered in this process and resolved to varying degrees. The following are examples:

- What are appropriate definitions for *sufficient suitable lands, market factor, appropriate urban densities, population figures, need for commercial land, need for industrial land, redevelopment potential?*
- How do/should cities or counties track residential densities, commercial development levels, and industrial development levels?
- What will be the impact of development applications—vested prior to the adoption of GMA comprehensive plans—on the development patterns (especially densities achieved) observed over the first five-year evaluation period?
- What measures are "reasonably likely to increase consistency" between planned and actual development?
- What are the minimum monitoring requirements for local governments?
- How should disputes between jurisdictions on issues related to land monitoring and buildable lands be resolved?
- How can the current and potential resources of cities, counties, and regional planning organizations be utilized most effectively to meet requirements of the statute?
- What resources (especially fiscal) will be needed to meet the requirements of the statute?

Conclusion

Much of the first two years following the program's creation was spent establishing the parameters for collecting and evaluating land supply data. In addition, local GIS land use data were updated and integration of other relevant information began. During the next phase, coordination will broaden and work will intensify. Results of the first evaluation, due in September 2002, will provide a comprehensive evaluation of Washington's land supply and growth management progress than ever before possible.

Source

Washington State Department of Community, Trade and Economic Development. 1999. *Buildable Lands Program Guidelines* (Draft). Olympia, Wash.: CTED, June.

CASE SUMMARY: CITY OF SEATTLE, WASHINGTON

Overview

The land monitoring and capacity analysis activities undertaken by the City of Seattle have occurred within a planning context shaped by Washington State's Growth Management Act (1990). The Act mandates local jurisdictions to prepare comprehensive plans that address, among other subjects, their ability to accommodate future population and employment growth within designated urban growth boundaries (UGBs). King County, which administers the UGB for the Seattle area, utilizes estimates of capacity provided by its constituent municipal jurisdictions to assess the overall adequacy of the land supply. Seattle completed a background land capacity study in 1991, in preparation for developing the city's Comprehensive Plan, "Toward a Sustainable Seattle." Land supply monitoring has been a useful tool in both areawide and neighborhood or "Urban Village" planning. In 1997 the city revised its methodology in response to work done by the King County Land Capacity Task Force and generated new capacity estimates.

Geographic Information System Resources

In the past, the city's analyses of development capacity have been heavily reliant on assessor's data provided by King County. At this time, the city has developed an integrated geographic information system (GIS) that utilizes more accurate zoning and parcel coverages, as well as in-house land use data in concert with housing units and floor area totals, and valuation data from the assessor.

Among the many problems associated with county assessor's data were the following:

- Inaccurate parcel area figures
- Unreliable valuation data that may be as much as a year out-of-date
- Inconsistent and unreliable address fields, particularly for use with the city's own address data (This field is not used by the city to track land supply or capacity.)

- Assessor's land use codes that are not updated to meet the needs of the city and are inconsistent with the land use planning classification schema
- Inaccurate zoning designations

To correct these and other problematic aspects of the assessor's data, including accuracy of the parcel attribute data and their compatibility with other databases, the city conducted multiple field checks and developed its own land use data in limited cases where they were deemed most needed for planning analysis. For example, a spot survey of industrial land use was conducted in 1997. In addition, the city now relies on its own GIS layers for data on parcel size and zoning.

The City of Seattle uses a combination of ArcInfo AML and SAS software to process data for capacity analyses. Individual parcel, tract, and sub-parcel records are extracted from tables produced by the GIS. Assessor's and other data are merged with these for analysis of building and land value relative to potential capacity.

The parcel layer itself is derived from data supplied by King County. Parcel boundaries are updated on a continuous daily basis. Parcel attribute changes have a six-to-eight-week lag time. Assessor's update cycles entail the following:

- For residential land use—physical inspection of each parcel and update of land use codes every six years, statistical update of assessed value for all on an annual basis.
- For commercial properties (including multifamily residential)—a somewhat higher percentage (20 to 40 percent) of properties are updated for land use annually.

Land Supply Monitoring and Capacity Analysis

The following methodology was employed in Seattle's most recent citywide capacity analysis (1997), which was an update of the 1991 analysis. The revised estimates incorporated new assumptions and methodological elements that differed from those in the original analysis, including higher residential densities in some commercial zones, lower densities in industrial zones, and updated assessment records to identify additional developable commercial land.

The analysis classified all parcels (or subparcel areas) within the city as follows:

- Vacant land—according to assessor's land use codes as well as valuation (parcels with an improvement-to-land-value ratio of .001 or less)
- Land available for redevelopment or infill—criteria vary by zoning (see the following paragraphs)
- Unavailable land—all publicly owned lands, plus rights of way, parks, cemeteries, and other land effectively precluded from additional development by current ownership or use

- Land within designated historic districts and "major institutions" planning areas (which include hospitals, college campuses, and areas surrounding them)

Two key determinants of capacity were considered:

1. Regulatory status—as determined by zoned land use and allowable density
2. Land use classification status—primarily vacant or underutilized

All vacant land was assumed to be available for development. Criteria for evaluating redevelopment and infill potential on non-vacant parcels varied by zoned land use type, as follows:

- Single-family infill potential was based on the presence of sufficient land (twice the minimum required lot area) on a parcel to allow for an additional house.
- Low-rise multifamily redevelopment potential was based on current development at 40 percent or less of maximum allowed units on a parcel. This criterion reflected observations of recent development activity in these zones. The city assumed full build-out upon redevelopment.
- For other multifamily zones (including commercial and mixed-use) and all parcels in the downtown area, redevelopment potential was based on an improvement-to-land-value ratio of 0.5 or less. The analysis assumed that no redevelopment would occur in industrial zones.

The reason for using different methods to determine redevelopment potential is that the low-rise multifamily zones have fixed density limits against which existing densities can be compared. The mid- and high-rise multifamily zones and the commercial zones, on the other hand, allow unlimited residential densities. To address the development potential of this latter category of parcels, the city needed a surrogate indicator of redevelopment.

Unless otherwise noted, residential, office, and commercial development was assumed to occur at the density limits set by zoning. Many zones that allow residential uses have no density limits, and the city used a weighted average of recently observed densities of new construction in those zones to calculate capacity. Industrial densities were estimated using limits imposed by building code requirements and information derived from observed development trends.

Because land in most commercial and downtown zones may be developed for residential, nonresidential, or a mixture of the two uses, three alternative scenarios were tested:

1. All available land in these zones developed as residential
2. New uses split between residential and nonresidential (in downtown the split varied by zone, but 50-50 applied elsewhere)
3. All available land in these zones developed as nonresidential

Although the second scenario was considered most likely to occur, addressing all three scenarios provided valuable information about the full range of potential development outcomes.

The City of Seattle applied a 15 percent market factor as a reduction from the calculated capacity for single-family and low-rise multifamily residential zones. For other types of zoning, consideration of the market was accounted for by estimating capacity based on a valuation ratio, assumed to be a valid market demand indicator. The analysis did not take into account factors such as owner preference, more refined market indicators, or specific site constraints that might limit the intensity (or even the possibility) of future development.

To assess the adequacy of the residential capacity to accommodate future population growth, the city employed a range of expected household sizes (1.7 to 2.1). The city also compared past development rates (2,000 units per year) with saturation over the 20-year planning horizon, as another measure of the adequacy of the land supply. In both cases it was determined that adequate development capacity did exist. Under the Growth Management Act (GMA) planning in King County, the City of Seattle has a 20-year growth target of 50,000 to 60,000 households. By agreement among King County jurisdictions, capacity should exceed the target set by at least 25 percent. With a capacity for more than 100,000 additional dwelling units, Seattle met that goal. The city also developed indicators (based on permit records) of how well it is meeting Comprehensive Plan goals for density of development.

The capacity analysis was performed for the city as a whole and for each of the Urban Centers and Urban Villages designated by the Comprehensive Plan. These areas, within which the majority of population and employment growth is planned to occur, form a key element of the land use plan for the city. Subarea estimates of capacity provided population and employment targets for subsequent neighborhood planning efforts for the Urban Centers and Villages. Currently, neighborhood-level capacity analysis is again being used to assess the impact of proposed rezones (submitted by the neighborhoods) on development capacity.

City planners indicated several problems that were encountered in conducting the analysis. First, as already noted, some zones did not impose specific density limits on development, and the city thus relied on land valuation ratios to identify redevelopable parcels. Second, demolition costs associated with redevelopment reduced the feasibility of development on some parcels. For this reason, the city employed density and valuation criteria that were restrictive enough to select only parcels where the return of redevelopment was likely to exceed the costs. (It should be noted also that the purpose of the analysis was to provide an overall estimate of how much growth the city can absorb, rather than predict the future development of individual parcels, thus lessening the importance of constraints associated with the potential redevelopment of particular properties.) Third, relating the zoning coverage to parcels was problematic. Because the boundaries of these two layers did not match, both silver polygons and split parcels resulted from a spatial overlay. For these reasons, the city applied a "50 percent rule": the zoning designation that covers over 50 percent of the parcel area was assigned to the entire parcel for the purposes of development capacity analysis.

Contacts

Carol Bason

Ellsie Crossman

Tom Hauger

Jennifer Pettyjohn

Strategic Planning Office

Sources

City of Seattle. 1991. *Background paper: Zoned development capacity.* Office of Long Range Planning, City of Seattle.

———. 1997. *Zoned development capacity.* March. City of Seattle.

CASE SUMMARY: SNOHOMISH COUNTY, WASHINGTON

Overview

Snohomish County is the northernmost county of the Central Puget Sound urbanized region. It contains a range of settings, from medium-size older industrial areas, to suburban residential and employment areas, to rural towns. It is a fast-growing county, with an expected population increase of more than 187,000 by the year 2012.

As mandated by the state's Growth Management Act (GMA) (1990), the county established Urban Growth Areas and countywide planning policies to provide a framework for county, subarea, and municipal comprehensive plans. In compliance with the Act, studies of residential and employment land supply were conducted in the mid-1990s to ascertain whether the proposed urban growth areas (UGAs) had adequate capacity for future growth. Countywide planning policies require an update of supply estimates and density assumptions every five years.

The methodology and data inputs for the land supply studies are summarized in the following sections, together representing a comprehensive inventory and analysis of the land supply within unincorporated areas of the county.

The county employed a parcel-point geographic information system (GIS) approach and applied multiple screens and capacity adjustment criteria that addressed environmental constraints, disaggregated demand characteristics, under-build factors, and land availability.

Geographic Information System Resources

Snohomish County's GIS was established in the early 1990s as a distributed system with shared database maintenance among multiple departments. The county is currently working on completing a GIS parcel polygon coverage, which will be available for use in the next major round of land supply updates. Working with the digitized parcels will enhance the county's ability to analyze the land supply by, for instance, maintaining an accurate record of parcel size and allowing for more accurate overlays of parcel with nonparcel data layers (e.g., wetlands, utility service-sheds) to determine levels of constraint on development. As a general monitoring instrument, the parcel coverage will provide the framework for a continuously updated inventory that will benefit a wide variety of users at little or no start-up cost to them and will provide benefits particularly to the assessment and planning departments.

Meanwhile, for the purposes of the capacity analysis done in 1993–1995, county planners worked with a parcel-point coverage, derived from parcel records geocoded by address. Standard address records were established for developed parcels for a recent 911 enhancement project. Undeveloped parcels were assigned "dummy" addresses for locational reference.

Assessor's records have been the primary source of parcel attributes used for land monitoring. In addition, in order to assign an appropriate planned use to each parcel, the county overlaid the digitized boundaries of land use plan designations with the geocoded parcels. Data sources for assessment of development constraints included the Snohomish County Stream and Wetland Survey, National Wetlands Inventory, Federal Emergency Management Agency (FEMA) maps, and United States Geological Survey (USGS) topographic and surficial geology maps.

Land Supply and Capacity Monitoring

Residential Land Capacity Analysis The *Urban Growth Area Residential Land Capacity Analysis* (1995) entailed a comparison of the 20-year supply and demand for land within the proposed unincorporated growth area. Individual cities in Snohomish County performed their own separate capacity analyses. The results of the cities' work were subsequently combined with county estimates to evaluate of the adequacy of the entire county urban growth area (UGA) to accommodate anticipated growth. Results of this analysis— comprising estimated "population holding capacity" for the areas under study—were used to support the 1997 adoption of the current urban growth boundary.

General assumptions guided the capacity analysis as follows:

- Within the UGA, a significant amount of land had been designated by pre-GMA comprehensive plan land use maps for nonurban uses—rural, residential estate, suburban agricultural, resource lands. With some exceptions, low-density urban residential uses were assumed to predominate over the 20-year analysis horizon.
- Under-build would continue to occur in new residential development, including single-family developments, where less-than-optimal platting reduces effective densities, as well as multifamily projects.
- Parcels on which there was already a nonresidential use were assumed to have no residential capacity.
- Parcels designated for commercial uses were assumed to have no residential capacity, even though multifamily development was allowed under current zoning in some of these areas.

Definitions of vacant, partially utilized, and underutilized land were based on the methodology established by the Washington State Department of Community, Trade, and Economic Development (CTED) (see Enger 1992). Density assumptions used in the analysis were derived from observed development trends and actual densities for the years 1990 to 1992.

The following summarizes the steps in the analysis:

1. Inventory residentially designated parcels, both vacant and developed.
2. Calculate theoretical yield for each parcel—acres multiplied by maximum units allowed per acre for each planned land use (truncated to whole units)—excluding existing easements and rights of way.
3. Subtract existing units from each parcel's theoretical yield.
4. Make further reductions to aggregate capacity to account for the following:
   a. Land not available over the 20-year horizon, accounted for by applying market factor deductions—15 percent for vacant land, 30 percent or more for partially utilized parcels (with higher deductions for subareas where such parcels under 1 acre in size exceeded 30 percent of total partially utilized acreage), and 30 percent or more for underutilized parcels (with higher deductions for subareas where the acreage of parcels for which the building value is greater than the land value exceeded 30 percent of the total under-utilized acreage). (These figures were based on CTED literature and surveys conducted by King County, Washington.)
   b. Public purpose lands—accounted for with a 15 percent deduction
   c. Critical areas, established as follows:
      - Streams and wetlands—50-foot buffers, representing the average required by existing county regulations (with some exceptions). The resulting acreage was subsequently increased by 13 percent to adjust for areawide underestimation of wetland areas.

- Geologically hazardous areas (landslide and seismic hazards)—excluding wetlands and stream buffers (to avoid double counting).
- Frequently flooded areas (if not in the first two categories).

   A composite map of critical areas was overlaid with the planned land use layer to obtain a percent figure for constrained lands for each land use type. An "encumbrance percentage" was also calculated as "an estimate of actual density reduction . . . due solely to critical areas," based on observed subdivision practices (e.g., Planned Residential Development [PRD] and lot size averaging) that effectively transfer residential density within sites. Multiplying these figures yielded a "critical areas reduction factor" for each county subarea.
   d. Future rights of way—accounted for with a 15 percent deduction for platted sites.
   e. Under-build, accounted for with a market factor deduction of 13.3 percent for single-family and 28.5 percent for multifamily uses. (Specific deductions represented the "residual density gap" between that allowed and that built during the 1990–1992 period.)
   f. Accessory apartment and duplex uses, accounted for with an adjustment for existing multiple units on single-family parcels.
5. Convert housing unit capacity into population capacity for comparison with future population growth forecasts. (This conversion employed occupancy rates and average household size from the 1990 Census for single-family and multifamily housing separately.)

The residential land supply analysis concluded that there was overall sufficient capacity to accommodate the forecast 20-year population increase (both exclusively within the unincorporated areas and for the full county UGA). In addition, of the supply of buildable lands, 58 percent were determined to be vacant, 35 percent partially used, and 7 percent underutilized.

Employment Land Capacity Analysis   The county's *Employment Land Capacity Analysis* (1995) utilized forecasts for employment land demand for three geographic subareas of Snohomish County: southwest county UGA, remainder county UGA, and non-UGA lands. Based on findings of a Snohomish County Business and Industrial Land Survey (1985), employment growth forecasts were disaggregated by sector and subsequently assigned to appropriate land use categories. The 1985 survey also provided a basis for establishing square feet of built space per employee, as well as a lot coverage factor for each employment sector type. Single-story construction was assumed in converting square feet of improvements to an acreage requirement.

   The demand analysis produced total acreage figures for the land required to accommodate job growth per land use, per sector, and within each county subarea.

   The analysis of the buildable land supply for employment-based land uses entailed conducting a disaggregated inventory of vacant, partially utilized, and underutilized parcels. Several criteria for inclusion in the inventory applied to all three categories:

- Designation by plan for employment use
- Not in government or church ownership
- Application of an appropriate critical areas reduction factor (as mentioned earlier)

The criterion for classifying parcels as vacant was absence of existing uses or improvements.

Criteria for classifying parcels as underutilized and determining total supply acreages included the following:

- Currently not vacant
- A building-to-land-assessed-value ratio of <0.5, implying that the structure was in poor condition, the parcel was zoned for a more intensive use, or the parcel was much bigger than that required to sustain its current use
- A 50 percent deduction for a "marketability and safety factor," to account for owners who would choose, for a variety of reasons, to hold onto the property in its current use without making further improvements
- Total supply discounted to account for current employment on underutilized parcels

Criteria for classifying parcels as partially developed and determining total supply acreages included the following:

- Currently not vacant
- A building-to-land-assessed-value ratio >0.5
- Parcels not considered "underutilized"
- Parcel area > 1 acre
- Exclusion of parcels with building coverage > or = 25 percent
- For lands currently in residential use with fewer than five units, a 50 percent deduction (or "safety factor") was applied to the supply total.
- Exclusion of parcels with existing multifamily residential uses with > four units
- Site coverage of existing employment uses, as determined from air photos (including buildings and paved areas, and other areas of the site in use)— parcels with >25 percent coverage were excluded from the inventory; for those with <25 percent coverage, only the vacant portion of the parcel was included in the supply.

The analysis concluded that the current supply of employment lands, when compared with employment targets, showed a deficit for some sectors in the southwest county subarea. The county subsequently used the supply and capacity findings to propose and evaluate a "diversified centers alternative," which transferred some of the future employment growth to outlying centers with adequate capacity.

In accordance with the advisory guidance of representatives from both the public and private sectors, future efforts to monitor and analyze non-residential land supply in the county will likely address the following:

- More detailed tracking of development trends and demand characteristics as a basis to adjust analysis assumptions
- A unified, full-county analysis, using data from both the county and the cities (especially GIS)
- Analysis of mixed-use urban centers
- Assessment of the impact of new regulations, especially higher allowable densities and minimum density zoning, on land supply and capacity

Contacts

Tim Koss, Senior Planner/Demographer
Stephen Toy, Principal Demographer
Planning and Development Services

Sources

Enger, Susan C. 1992. *Issues in designating urban growth areas.* Olympia, Wash: State of Washington, Department of Community Development, Growth Management Division.

Snohomish County Planning and Development Services. 1995. *Employment land capacity analysis for unincorporated Snohomish County.* June revision.

———. 1995. *Urban growth area residential land capacity analysis.* Summer.

CASE SUMMARY: METROPOLITAN COUNCIL, MINNEAPOLIS–ST. PAUL, MINNESOTA

Overview

The Metropolitan Council was created in 1967 as a regional governing body for the seven counties that encompass the urbanized areas of Minnesota's Twin Cities, Minneapolis and St. Paul. These seven counties include 189 local jurisdictions. The region currently has a population of more than 2.5 million, which is expected to grow to 3.1 million by the year 2020.

The Twin Cities region is characterized by low-density development and, outside the Sun Belt, is second only to the central Puget Sound (Seattle) in regional growth rate (10 percent population increase between 1990 and 1997). Loss of farmland, the need to maintain water quality, and the concentration of poverty in the central cities have been major concerns of regional planning and policy-making. Of particular concern are the cost of infrastructure to serve sprawling development patterns and the ability to provide affordable housing within suburban areas.

A designated Metropolitan Urban Service Area (MUSA) has been in place since the 1970s, delineating the area within which the Metropolitan Council will support urban growth through the year 2020 by providing adequate urban services and infrastructure. Regional planning for the MUSA, as laid out in the *Regional Blueprint* (1996a), is done in concert with local jurisdictions' comprehensive planning efforts. Specifically, long-term growth planning for subregional areas is implemented through the designation of "urban staging areas," which are established through local plans. The staged-growth concept is intended to ensure sufficient developable land for a range of land uses, adequate levels of affordable housing, and the prevention of leapfrog development. Local communities set housing density benchmarks in cooperation with the Council, which also provides targeted funds and infrastructure investments to assist with plan implementation, and technical assistance to facilitate coordinated planning for the region.

The land supply monitoring work that is undertaken and overseen by the Council provides a significant example of interjurisdictional coordination within the context of a strong regional governance structure. Although the regional analysis is not parcel-based, the Council has developed a sophisticated approach that addresses multiple time frames (staged growth), subarea-specific demand assumptions, and input from multiple local jurisdictions.

Geographic Information System Resources

The Metropolitan Council initiated the MetroGIS project in 1995. The purpose of MetroGIS is to provide an organizational framework and resources for geographic data sharing between local governments and private and public organizations in the region. Among MetroGIS's primary objectives is the development and maintenance of regional data sets and the provision of technical and organizational assistance to member agencies and organizations. Interjurisdictional database issues, such as inconsistent land use classifications and zoning codes, have hindered the integration of geographic information systems (GIS) regionally. A negotiation process to establish regional data sharing agreements is currently under way.

The Council is also working to develop parcel-level GIS capability, with the goal of eventually constructing a unified system with full regional coverage. The implementation of parcel-based GIS has occurred piecemeal among the region's subareas and jurisdictions. However, a MetroGIS survey of local governments revealed that many are in the process or planning stages of developing a parcel layer for county and municipal GIS. For most current applications using mapped parcel data, the Council employs assessor's records linked to geocoded parcel points rather than to digitized parcel polygons.

The Council mandates that local jurisdictions perform five-year land use inventories. Either local jurisdiction or MetroGIS data are used in these efforts. The comparability of land use data is aided by the conversion of local data to generalized land use classifications (the Council provides guidelines for conversion of local classifications to a regionally consistent set of definitions). The Council also collects data on new construction for each jurisdiction on an annual basis.

Land Supply and Capacity Monitoring

Regional Land Use Inventory The Metropolitan Council updates a regional land use map every five years. Land is classified according to 14 general categories that apply regionwide. The land use coverage is generated through a hybrid process involving primary interpretation of aerial photos, cross-checked with GIS data, some of which is at the parcel level. Land use data are also updated from permit records supplied by the counties. The resulting land use coverage is composed of variable-sized polygons of contiguous land use.

Community Growth Allocations The Metropolitan Council develops regional growth forecasts for both employment and households. The disaggregation of these forecasts is thereafter made to subregional planning areas— primarily individual jurisdictions. This allocation process relies on estimates of land supply and capacity in order to adjust subregional growth forecasts appropriately.

Future amounts and densities of development are derived from analyses incorporating data on subregional land development trends, existing land use mixes, and local land development regulations. The characteristics of the demand for residential land and housing within each local area are also adjusted to reflect demographic trends (e.g., age and size of households).

Assumptions that guide the sub-regional allocation process include the following:

- Significant amounts of redevelopment and reinvestment will occur in the central cities.
- Older developed ("inner-ring") suburbs will achieve build-out (or near build-out) as infill occurs on vacant parcels.
- A recommended three dwelling units per acre minimum for newly developing communities.
- A standard of one dwelling unit per 40 acres in permanent agricultural areas and one dwelling unit per 10 acres in designated rural areas.

The demand for nonresidential land is also addressed, although the relationship between employment growth and land consumption is somewhat less linear. For example, as noted by Council planners, a significant amount of employment growth in the region has been absorbed within existing facilities and already developed sites. This phenomenon makes the estimation of future employment-generating land uses problematic, and it reflects the widespread problem, especially in suburban areas, of overzoning for nonresidential land uses.

The Metropolitan Council methodology for calculating capacity for the region as a whole does not account for infill or redevelopment potential. (Without historic data on land conversion at the parcel level on which to base screening criteria, this type of analysis has been judged not feasible.) However, the Council does work with local communities to identify lands for potential redevelopment, with the particular aim of achieving higher densities along transportation corridors as well as within the transportation nodes and employment centers that have been targeted for transit-oriented development. The Metropolitan Council has issued guidelines for the identification of likely redevelopment areas, focusing on tax-forfeited properties, brownfields sites, obsolete retail sites, areas with declining property values, and locations near new retail, transit, or other services amenities.

Subregional growth figures are revised in collaboration with local planning officials to reflect local conditions. During the most recent update of the Community Growth Allocations, local input resulted in a one-third downward adjustment to aggregate regional land capacity estimates—primarily because of the deduction of land dedicated to parks and open space, "unavailable" private parcels, and land erroneously included in the inventory based on inaccurate permit data.

Planning and Analysis for the MUSA and Urban Staging Areas The Council-delineated MUSA indicates the maximum extent of urban services and infrastructure planned through the year 2020. The Urban Reserve Boundary encompasses eventual serviced land supply through at least the year 2040. The boundaries partly reflect watershed areas and partly jurisdiction boundaries. Estimates of the capacity of the land supply within these areas to meet future demand are based on historical land use mixes, patterns, and densities. The Metropolitan Council encourages the use of GIS by local planners (where available) to perform necessary analyses, and continues to work to improve the GIS resources at its own and local jurisdictions' disposal. Federally mapped wetlands have been applied as a constraint to developable lands in most areas.

Delineation of the Urban Staging Areas that constitute the MUSA involved identifying those parts of the Urban Reserve to be converted to urban uses through the year 2020. Local governments are also required to plan for a post-2020 "holding zone" for future urban development and long-term planned extension of urban services. Land use consumption and infrastructure demand in these areas will be monitored in the interim.

Approval of any expansion of the MUSA requires agreement between the local jurisdictions and the Council about existing and future land supplies, anticipated demand, and additional land needs implied by comparing the two. Requests by local governments for MUSA expansion must have supporting documentation that relies heavily on measured land supply needs. Monitoring of development patterns and densities will continue as a follow-up to any expansion.

In addition, a set of "Performance Indicators" are monitored at the regional level, which include the following:

- Degree of consistency between actual and planned land uses
- Increases in densities within the urban service areas
- Reductions in the rate of development of "prime farmland"

According to Council staff, political and market dynamics have stymied past efforts to match demand with supply forecasts at the local level. Data and analysis capabilities provided by a regional parcel-based GIS (under development) would improve this situation, and it is expected that with greater uniformity among all jurisdictions (and other interested parties) in having access to high-quality, high-resolution GIS data, disagreements about capacity will decrease markedly.

Design Center for American Urban Landscapes: North Metro I-35W Corridor Comprehensive Livable Community Urban Design Framework and Transportation Study

In a related initiative, MetroGIS provided funding to the I-35W Corridor Coalition, a group of seven cities and two counties working to foster more effective transportation, economic, and community development planning within the corridor of a major interstate highway running through the region. This work is being conducted with assistance from the Design Center for American Urban Landscapes at the University of Minnesota.

The study area for the project encompasses 77 square miles, 151,000 residents, 83,000 jobs, and 46,000 parcels of land. The communities in the area are characterized as "inner-ring" and "second-ring" suburbs, with good transportation connections to the central cities. A significant amount of new commercial construction has occurred in recent years, both on new sites and through the development of brownfields sites.

A primary initial objective of the Coalition is to develop parcel-level land use and planned land use GIS coverages that will serve as resources for subsequent analyses of land supply, development, and redevelopment potential. According to MetroGIS, this project may provide a "model for further integration of land use maps throughout the metropolitan region."

The GIS development effort is in its early stages, focusing on a range of database development issues, which include the following:

- Standardizing land use and plan designations and parcel coding schemes among the participating jurisdictions
- Linking county assessor's data with buildings, addresses, and parcels
- Converting tax assessor's data into usable land use planning data
- Breaking down the land use data into meaningful subregional districts
- Obtaining data collected by private firms on industrial and commercial land uses and efficiently converting them to a usable format

The study area has a large amount of developable land. A primary objective of the Coalition's work is to capitalize on the economic development and redevelopment potential of the land supply. It is anticipated that the parcel-based GIS will be an essential tool to

- identify individual development sites;
- assess the impact of development on transportation demand;
- market available developable lands to developers; and
- establish interjurisdictional consistency in land monitoring procedures and definitions.

Transportation planning is another major focus of the project, with particular emphasis on designing a land information system to explore the interaction between built form, residential densities, and land development within major transportation corridors. Project leaders have decided that such work requires disaggregating the land information down to the parcel level.

Contacts

Bob Davis, Senior Planner
Richard Gelbmann, GIS Coordinator
Michael Munson, Senior Planner
Metropolitan Council
Carol Swenson
Design Center for American Urban Landscapes

Sources

MetroGIS. Official web site. Available August 1998 at [http://www.metrogis. org].

Metropolitan Council. 1996a. *Regional blueprint: Twin Cities metropolitan area.*

————. 1996b. *Local planning handbook.*

————. Official web site. Available November 1997 at [http://www. metrocouncil.org].

North Metro I-35W Corridor Coalition. Official web site. Available July 1997 and November 1998 at [http://elsie.hamline.edu/135W/index.html].

Welsh, Dennis, David Windle, and Jerry Hopkins. 1997. North Metro I-35W Corridor Coalition: A work in progress. Documented presentation to the MN GIS/LIS Consortium, October 3.

CASE SUMMARY: CAPE COD COMMISSION, MASSACHUSETTS

Overview

The Cape Cod Commission is the regional planning organization for the 15 towns that constitute the Cape Cod region of Massachusetts. The Commission's planning activities focus on environmental protection, water quality, and transportation, as well as on the impacts of current and future land use. The *Regional Policy Plan* (1996) describes concerns about growth and concomitant land use changes on the Cape as a driving force behind the creation of the Commission itself in 1990. Problems persist today regarding the preservation of sources of drinking water, fragile natural areas, and the Cape's cultural heritage, which is reflected in the rural landscapes and small town character of its settlements.

The recent land monitoring and analysis work of the Commission described in the case summary focused on quantifying the potential impacts of growth on natural and man-made systems. Several recently published "capacity" studies—here defined as "carrying capacity"—addressed the possibility that future growth would exceed acceptable thresholds for these systems. The studies utilized "build-out" analyses (described in Chapter 2 as "maximum" development capacity) to estimate potential future land use changes and various location-specific impacts of new development. The Commission also attempted to capture the effects of specific development patterns (e.g., residential sprawl and strip commercial).

Geographic Information System Resources

The Cape Cod Commission's geographic information system (GIS) contains a regionwide layer of digitized parcels, which was nearly complete at the time the studies described herein were performed. Parcel attributes are supplied by the assessor's office. Land use codes (based on McConnell land cover classifications) assigned to the parcels by the assessor were deemed reasonably accurate for the planning applications of the GIS.

Additional GIS data layers include roadways, wetlands, aquifer recharge areas, public wells, forest cover, critical upland areas, roads and highways, zoning, Transportation Analysis Zones (TAZs), census geography, and water bodies—all of which were used in the build-out and carrying capacity analyses.

Land Supply and Capacity Monitoring

Two recent studies by the Commission applied land supply and capacity analysis to assess the future impacts of growth and development on the natural and man-made systems. Build-out forecasts were conducted for a series of future time horizons (2005, 2010, 2015, and 2020), as well as for "full" build-out, with all available lands developed to the limits set by prevailing regulations. Interim forecasts were based on a variety of data, including baseline land use and supply, recent observed rates of growth and development, census data, and local area-specific assumptions about the future absorption of remaining zoned capacity.

Outer Cape Capacity Study (1996) The *Outer Cape Capacity Study* evaluated "the impacts of growth on the municipal finances, public facilities, water supply, road system, open space, and ecology" of the Outer Cape subregion. It estimated the impacts of future development of residences and commercial establishments on water supply, roads, natural areas, local government services, and fiscal health of the local jurisdictions. The study emphasized residential land uses both because they were predominant in the study area and because, in quality and availability, data about residential supply and demand exceeded those pertaining to other types of uses.

The study model generated growth projections for winter and summer populations aggregated to Transportation Analysis Zone (TAZ) and Census Block Group (CBG). Study assumptions about future demand for land were based on data from the 1990 Census, the Commission's inventories of lodging units, Census figures on average household size per town, residential growth rates for towns based on past building permit records, and estimates of full build-out of residential land in each TAZ, considering the amount of undeveloped and under developed land as well as local land use regulations. Estimation of future commercial development was based on the amount of existing "business" lands or developed and potentially developable business lands (zoned for such use) within each jurisdiction.

For the build-out analysis all supply data were collected and categorized by CBG, with the exception of the jurisdiction of Wellfleet, where more detailed data were available. The analysis comprised the following steps.

Step 1: Each parcel record within the study area was coded by its CBG. All vacant parcels were selected, as well as large single-family parcels (>5 acres). For each CBG, estimates were made of the number of potential new residential parcels that could be produced through subdivision of selected parcels. This calculation employed jurisdiction-specific factors such as minimum lot size allowed, subdivision standards, timing of future subdivisions, rights-of-way discounts, and density thresholds for identifying infill potential on larger parcels with existing development. The results yielded a number of developable parcels, total acreage, average parcel size, and distribution of the parcels by size within each CBG.

Step 2: The inventory of lands generated by Step 1 provided the base for estimating build-out, both residential and commercial. As with the land supply estimates, each local jurisdiction provided local assumptions to guide the calculation of expected build-out. Such input was especially important for estimating the capacity of new commercial development, which varied significantly by locality.

Land supply figures originally aggregated to CBGs were further aggregated to each TAZ, and an expected growth rate (derived from data on past development) was computed for each TAZ proportionally, according to the amount of vacant or other developable land present.

Rough estimates of land consumption per employee were made based on current nonresidential land use patterns. It was assumed that the relationship between number of employees and land needs would not change significantly during the next 20 years.

The overall analysis generated TAZ-level build-out totals for both residential and nonresidential land uses for each of the horizon years. Totals included units constructed, land consumed by new development, commercial square footage and remaining capacity, and rough estimates of when full build-out would be achieved in each land use category. Lacking information on site-specific development constraints, the growth estimates for developable lands produced by this study were considered an upper limit of actual build-out capacity.

Monomoy Capacity Study   A companion study, the Monomoy Capacity Study (Cape Cod Commission 1996), was performed for the lower half of the Cape Cod region. Build-out analysis assessed the development potential of individual parcels within the study area, and the results provided input to the overall carrying capacity model for the subregion. The study's determination of development potential at the parcel level allowed for flexible aggregation to the zoning district, census tract, and town levels of geography. Consideration of zoning in relation to current land use was the primary means of determining development potential for both residential and nonresidential uses. (The study's build-out analysis did not attempt to look at market conditions or any other indicator of demand.)

The GIS parcel layer was overlaid with the layers for wetlands, roadways, and water bodies to identify constrained lands for subtraction from the inventory of available supply. Notably, because of the different scales of the data layers combined through overlay (e.g., parcels at 1:600 and wetlands at 1:25,000 scale), determinations of the developability of individual parcels were inconclusive in many cases.

Using the unconstrained parcels as a base, build-out totals for the years 2005, 2010, and 2015 were computed. These drew on anticipated annual growth rates for residential development based on the median number of housing units constructed each year during the period 1985 to 1994. Anticipated growth rates for industrial and commercial uses were based on regional employment forecasts. The study projected local increases in industrial and commercial square footage through a proportional allocation of regional growth in overall employment. A comparison of longitudinal data for industrial and commercial construction over the period 1985 to 1994 against the annual mean employment increase for the same period found these relationships to hold for the historical period, thus validating the allocation assumptions.

The build-out analysis entailed the following steps:

Step 1: Establishing categories of development potential. The land supply was divided into eight categories of use and developability: existing single-family residential, existing multifamily, developable residential, existing commercial, developable commercial, existing industrial, developable industrial, and undevelopable land. The classification of these lands relied on the following set of assumptions about future land use change:

- Unless otherwise specified, municipal land was withheld from the inventory.
- Existing commercial uses would remain commercial regardless of zoning.
- All residential uses currently in commercial or industrial zones would redevelop.
- Undeveloped commercial lots in commercial zones would develop as commercial uses.
- "Land-rich" commercial uses (e.g., golf courses) and unrestricted agricultural uses were likely to face future redevelopment pressure.
- Multifamily residential developments of (three) or more units were considered "highest and best uses," thus precluding redevelopment.

Step 2: Establishing development potential formulas. In residential zones, parcels with more than twice the minimum lot size (with 15 percent deducted for roads) were considered subdividable, and the rest assumed to develop to the density set by zoning. Minimum lot sizes were based on zoning or the accepted practice in each jurisdiction. Lots below the minimum were considered undevelopable.

Floor area ratio (FAR) was the standard for computing the potential for future development on land zoned for commercial and industrial uses.

Parcels identified as having development potential were aggregated to the census tract level.

Step 3: Commercial and industrial build-out. As established through Steps 1 and 2, the buildable land supply in commercial and industrial zones contained both vacant developable parcels and developed parcels with some capacity for further development or redevelopment. A formula was derived for each zoning category to establish a standard allowable FAR for parcels within each zone.

Step 4: Residential build-out potential. The analysis of residential build-out focused on single-family development; it was assumed that little multifamily development was likely to occur. To convert build-out figures into projections of year-round population, total units were multiplied by the number of persons per household. Further, several scenarios were generated for population growth to account for variable amounts of conversion from seasonal to year-round residence on the Cape.

The study concluded that full residential build-out is expected to occur long before commercial or industrial build-out in all of the subareas . Further more, the study went on to state, build-out presents a "fictional future" that is useful "as a point of departure for studying the carrying capacity of the region." The build-out findings point to "potential pressure points created by current zoning allowances . . . and differences between towns." Finally, discrepancies noted between build-out potential and amount of undeveloped land in an area can serve to identify and address problems with the current zoning.

Cape Cod Industrial Land Survey (1997)   The *Cape Cod Industrial Land Survey* of 1997 updated inventories of industrial lands conducted in 1992 and 1994. Data sources for classifying the industrial land supply included local zoning maps, aerial photos, local officials, the *Massachusetts Natural Heritage Atlas, the Regional Policy Plan Atlas,* the Census, and the Cape Cod Commission GIS. The unit of analysis for the inventory was the "industrial area" (subareas of contiguous industrial lands). Each industrial area was coded for various attributes, including vacant acres; fiber-optic infrastructure; water, electric, gas, and sewer or septic service levels; road class; land use; and zoning.

The survey found that 1,300 of 3,700 total acres of industrially zoned land were vacant and available for industrial development. GIS overlay analysis indicated that a majority of the parcels identified lay wholly or partly within public well supply zones, indicating unsuitability for additional industrial uses. However, the analysis did not identify or quantify the portion of individual parcels lying within the well supply zones—in other words, the overlay did not split the parcels—resulting in an aggregate overestimate of constrained industrial lands. The survey report emphasized that the level of accuracy of the analysis was appropriate only for the purposes of initial screening.

Contacts

Gary Prahm, Systems Manager for GIS
Cape Cod Commission

Sources

Cape Cod Commission. 1996. *Monomoy capacity study.* July.
———. 1996. *Outer Cape capacity study.* December.

————. 1996. *Regional policy plan.* November.

————. 1997. *Cape Cod Industrial Land Survey.* 4th ed. February. <http:www. capecodcommission.org/ilsurvey/ilsurvey.htm> (July 1, 1998).

CASE SUMMARY: MUNICIPALITY OF ANCHORAGE, ALASKA

Overview

The Municipality of Anchorage (MOA) is the jurisdiction that includes the greater metropolitan area of Anchorage, Alaska. After a spike of intense population growth during the early 1980s, followed by a steep decline, the region is now experiencing moderate population growth and increase in construction activity. Recent land use planning efforts have focused on the Anchorage Bowl, the 100-square-mile area constituting the heart of the municipality. Shaped by Alaska's cyclical economy over the past 25 years, land use patterns in the Anchorage Bowl are uneven, with many vacant and underutilized parcels, especially in the proximity of the Anchorage Central Business District. Analyses related to land supply and capacity were recently performed as part of the ongoing comprehensive planning process.

Geographic Information System Resources

The MOA implemented its first geographic information system (GIS) in the early 1980s. The parcel coverage was first operational in 1982. Although system development was initiated by the planning department, it is now maintained as an enterprise-wide GIS, serving multiple departments within the municipal government. Public works maintains the parcel layer, and other departments maintain other layers according to their departmental functions.

The land use database was compiled in 1994 in preparation for the Anchorage Bowl Comprehensive Plan Update. This effort coincided with a restructuring of the land use coding scheme (a hierarchical four-digit coding system developed in-house). The scheme replaced tax assessor's land use codes, which were based on structure type and lacked functional land use data necessary for planning purposes. Nonresidential land uses were classified through windshield survey supplemented by other sources of land use information, including the Yellow Pages, an area street directory, the city directory, and aerial photography. Residential land uses were classified by the Department of Community Planning and Development.

In conformance with the new land use coding scheme, parcel attributes included primary as well as secondary uses (as determined by relative percentage of lot coverage or extent of improvements). MOA planners used "special GIS lines" that effectively split parcels to delineate distinctive use areas (especially as land uses occurred on portions of large, mostly vacant parcels). Assessor's records provided data on parcel ownership and assessed value of land and improvements. GIS layers essential to the analyses included land use, zoning, environmental features (see the description of such analysis in the following section), water and sewer lines, streets, and planning subareas.

Recent database development efforts included an inventory of residential addresses for the entire municipality, an update of residential land use classifications, and a survey of land use changes since the last inventory in 1994, which utilized the building permits layer and aerial photographs. After completion of these projects, MOA planners planned to redo the land supply inventory and analysis, taking advantage of improved linkages between appraisal, permitting, and land use data files within the GIS.

Land Supply and Capacity Monitoring

The Department of Community Planning and Development carried out the Anchorage Bowl Vacant Land Suitability Analysis in 1997. This analysis involved parcel-specific identification of vacant lands, determination of suitability of these lands using overlays of environmental constraints, and aggregate analysis of total suitable lands by zoning category and planning subarea.

Environmental layers, overlaid on parcels to estimate degree of constraint on development, included wetlands (classes A, B, and C), avalanche hazard areas, slopes, floodplains, seismic hazard areas, and alpine areas. For parcels with more than one constraining factor present, the most constraining of these was used to determine developability.

Suitability criteria for development of vacant lands were as follows:

- *Generally suitable:* All areas not listed as "marginally suitable"or "generally unsuitable" (see the following criteria).
- *Marginally suitable:* Parcels containing moderate avalanche hazard areas, slopes of 16 to 35 percent, classes B and C wetlands, 100-year floodplains, or seismic hazard areas (zone 4).
- *Generally unsuitable:* Parcels containing high avalanche hazard areas, slopes greater than 35 percent, floodways, class A wetlands, seismic hazard areas (zone 5), or alpine areas (above the tree line—generally more than 2,000 feet in elevation).

Ownership records provided data to identify adjacent or nearby parcels in common ownership, as those are likely to be used for parking or other uses associated with the "primary" parcel.

Municipal planners used 1980 land use maps in conjunction with 1994 land use data to develop GIS maps showing land consumption from 1980 to 1994. Parcels developed since 1980 were identified by their 1994 land use. The Department of Community Planning and Development plans to compile data on land absorption from 1994 to the present, the results of which will be used to project build-out of the land supply from observed trends, as well as to provide input for a new transportation model.

HDR Alaska, Inc., a private consulting firm, recently examined industrial land supply in Anchorage. Study methods and results were published in the *Commercial and Industrial Land Use Study* (1996). To prepare data for the study, HDR merged the boundaries for five distinct analysis areas (referred to as "study units") with data provided by the MOA, which yielded a complete coverage of parcels coded for zoning and land use. The land supply inventory completed for this study considered only the primary uses on each parcel (despite the fact that secondary land use codes also indicated many significant subparcel land uses). These data comprised Database 1 and included more than 66,000 parcel records.

The development of Database 2 entailed the following:

- Selection of parcels with commercial or industrial uses as well as those that were vacant but in commercial and industrial zones, for a total of more than 16,000 parcel records.
- Linkage of selected parcels with assessor's data files (CAMA database). Parcel attributes in the CAMA database included building square footage, physical condition, assessed value, and parking. (The study report notes considerable problems with the CAMA data—attribute errors, missing PIN numbers, and out-dated parcel boundaries. However, because this was a "macrolevel" study, parcel-specific accuracy was not considered crucial.)
- Analysis of selected parcels in ArcView via queries and overlays with environmental and infrastructure data.

Finally, database development and initial analysis yielded the following:

- Existing land uses tallied by zoning district within each study area.
- An inventory of improvements (building square footage per parcel) for each land use type by study unit.
- Acreage and parcel size summarized by zoning category.
- Two measures of building condition: assessor's rating and assessed improvement value by square footage of building.
- Vacant land supply—total acres and parcel sizes per zoning district for each study unit.
- Redevelopable land supply—three categories derived from improvement/land value ratios: (1) high redevelopment potential (ratio of 0—coded as in use, but having no improvement value, e.g., parking lots), (2) moderate redevelopment potential (ratio between 0 and 1), and (3) low redevelopment potential (ratio between 1 and 2). Parcels with a ratio higher than 2 were considered to be fully developed and were not further considered in the redevelopment analysis.

The analysis excluded government-owned or other tax-exempt land.

(The report notes that, overall, the methods used may be less accurate for industrial lands than for residential or commercial, largely because fully developed uses may not have entailed actual structural improvements recorded by the assessor.)

Analysis of environmental constraints to future development considered exclusively vacant lands. Specific constraints included "preservation" wetlands, "conservation" wetlands, 100-year floodplains, floodways, moderately steep slopes, steep slopes, and high seismic hazard areas. A polygon overlay was performed in ArcView to select all parcels that overlapped (in part or entirely) with each of the constraint polygons—the entire acreage of those parcels was thus identified as "affected" by the constraint. This limited the ability of the analysis to identifying the proportion of parcels constrained by these factors. The analysis was also limited in its ability to take into account the cumulative impact of multiple constraints on individual parcels.

Determining "serviceability" of vacant land considered sewer and water utility servicing. This analysis relied primarily on assessor's records as to service levels of each parcel. Digital data on the location of sewer and water lines were also used to create a buffer of 500 feet to the boundaries of vacant parcels (once again, because of software limitations, identifying only those parcels that intersected with the buffered areas).

Measurements of site accessibility employed several criteria, including assessor's rating of accessibility (based on street features and traffic volumes), proximity to streets in the official street plan, as well as proximity to major transportation infrastructure (e.g., airport, rail lines). These measures included selecting parcels that fell within buffers of 100, 200, 300, 400, and 500 feet from each highway and road class. The information gained from this analysis guided a further detailed, use-specific analysis of site constraints relative to transportation accessibility.

The study also incorporated information gathered from interviews with "technical resource contacts" in the community—consisting of individuals with recognized expertise in the local economy and commercial and industrial real estate markets.

Contacts

Fred Carpenter, Technical Services Supervisor
David Tremont, Senior Planner
Department of Community Planning and Development

Sources

Municipality of Anchorage. 1996. *Anchorage Bowl comprehensive plan: GIS land use codes*. Department of Community Planning and Development.

———. Official web site. Available November 7, 1997 at [http://www.ci.anchorage. ak.us].

HDR Alaska, Inc. 1996. *Anchorage Bowl commercial and industrial land study*. Prepared for Municipality of Anchorage, Department of Community Planning and Development.

CASE SUMMARY: SAN DIEGO ASSOCIATION OF GOVERNMENTS, CALIFORNIA

Overview

The San Diego Association of Governments (SANDAG) is the metropolitan planning organization for San Diego County and the jurisdictions that lie within it, including the City of San Diego. The county is home to some 2.8 million residents and covers an area of more than 4,200 square miles, comprising a wide range of types and intensities of land use. Land capacity analysis by SANDAG does not represent an application of a parcel-based geographic information systems (GIS) (as described in this book). However, SANDAG is regarded as one of the strongest regional GIS sites in the country. The analysis of regional growth described here is a significant example of the use of GIS for modeling land consumption and regional development, one that integrates small-scale subregional capacity estimates with land use and transportation modeling at a regional and subregional level.

Geographic Information System Resources

The Regional Urban Information System (RUIS) is a joint city-county project and the region's multiparticipant GIS. Its development began in 1984. SanGIS, an independent public agency dedicated to joint data management and access to countywide GIS data, coordinates RUIS for SANDAG and various local governments. The SanGIS database contains more than 200 layers of geographic information, including a complete parcel coverage for San Diego County. The parcel coverage was originally digitized from assessor's base maps and continues to be maintained by the county assessor's office. SanGIS oversees the process whereby participating local government agencies maintain the data layers specific to their own activities and geographic areas. The system is structured as a distributed network linking multiple participants.

Data layers in the regional GIS include parcels, lots, easements, topography, various municipal and tax districts, hazard areas (geologic, earthquake), floodplains, and utilities.

Remote sensing (both satellite imagery and, more recently, digital aerial photography) is used to update a land use database for the region. This was considered to be a cost-efficient approach to large area classification and mapping. For the 1990 inventory, land uses were classified for areas of contiguous homogeneous land use as small as 2.5 acres. For the 1995 land use inventory, uses were assigned to individual parcels through map overlay with the

SanGIS lot and parcel polygons. Although this most recent effort is not a true parcel-based inventory, SANDAG is moving in that direction. The agency is compiling a regional land use map with data derived from the following:

- The 1995 generalized land use map (based on 1995 aerial photos and satellite images)
- Constrained lands (steep slopes, flood areas, public lands, wetlands, airport noise contours, and future freeway corridors)—identified and delineated using GIS coverages and refined with the assistance of local planners
- Community land use plans (using four-digit hierarchical classification codes), which were interpreted in consultation with local planners, who provided clarification as to maximum numbers of units allowed or expected according to the plan, as well as the identification of multiple use and special plan areas
- Community transportation plans
- Potential redevelopment areas
- Site-level information on large-scale projects that are either in process of development or recently completed

To increase the accuracy of the mapping efforts, SANDAG produced and distributed to planners at local jurisdictions a series of maps to solicit their input. Thematic maps included community plans, circulation elements, existing land uses, potential redevelopment sites, constraints to development, and site-specific projects.

SanGIS is currently improving the accuracy of the assessor's parcel numbers in its parcel boundaries map layer. The land use codes in the assessor's Master Property Record File are not regularly maintained and are of limited use for land use planning analysis. (Updating of assessor's files in California has been performed in frequently since the passage of Proposition 13, which significantly reduced funding for such work.) SanGIS is working to develop its own parcel-level land use coverage by coding parcels through an overlay with air-photo-derived land use polygons, checked against information from local planning departments.

Land Supply and Capacity Monitoring

SANDAG administers the Regional Growth Management Strategy, a coordinated effort to manage the effects of growth on the quality of life in San Diego County (e.g., on air quality, transportation, open space, and housing). The Regional Growth Forecast, a periodic effort, has been a primary element of the Growth Management Strategy since the early-1970s. As described in the following section, land use data and capacity estimates are employed as inputs to the subregional allocation models that generate forecasts for population and employment growth.

2020 Regional Growth Forecast (1998) The Regional Growth Forecast provided regionwide totals of population, housing, and employment for the period 1995 to 2020. SANDAG utilized the Urban Development Model (UDM) to allocate regional growth forecasts to subregional areas. UDM first disaggregated population and employment forecasts to 208 Zones for Urban Modeling (ZUMs), which are composed of aggregated census tracts. UDM then further distributed the ZUM forecast to the smallest set of land units, the Master Geographic Reference Areas (MGRAs), which were generated through an overlay of census blocks (some split by topographic features), census tracts, community planning areas, municipal boundaries, spheres of influence, and zip codes—producing 25,354 of these zones. (Parcel boundaries were incorporated with the model output, but not utilized as a geographic layer for analysis.)

The allocation of residential and employment development to ZUMs was based on a standard spatial interaction model that accounted for land use, population, employment distributions, and accessibility between zones. The allocation process was sensitive to limits in the development capacity within ZUMs, which were calculated as a function of the amount and type of vacant land present, current residential density, and estimated redevelopment and infill potential.

The suballocation of ZUM-level forecasts to MGRAs was based on accessibility to existing residential and employment uses. This process was also sensitive to land supply and development capacity limits. Model equations accounted for land uses and housing units gained and lost through development and redevelopment.

At both levels of disaggregation, allocated growth that exceeded current development capacity was reallocated to nearby analysis zones (measured by accessibility) that did have adequate capacity.

As households and employment were allocated to small areas, UDM predicted and accounted for changes in land use, starting from an established baseline land supply. Land use change adhered to the following assumptions:

- Employment growth was assumed to occur through development of vacant land for industrial, commercial/service, office, and school use as well as through residential redevelopment, and infill.
- Multifamily development was assumed to occur through development of vacant land and redevelopment of single-family or mobile home areas, and infill.
- Single-family development was assumed to occur through development of vacant low-density residential land (<1 dwelling unit/acre) and of vacant higher-density single-family land (>1 dwelling unit/acre), as well as through mobile home redevelopment, and infill.

Preparation of land supply data for use in regional models proceeded as follows.

SANDAG utilized the regional land use coverage in concert with other GIS layers to identify vacant lands. Also identified were lands with a potential to change use through redevelopment, as well as lands with infill potential. The latter were determined based on differences revealed through GIS overlay between land use plan designations and current land uses and intensities. Information from local planners corrected classification errors and allowed SANDAG to remove from the inventory some lands deemed not developable for a variety of reasons that the regional analysis had not screened for, such as viable agricultural and extractive land uses. The end product was a land supply base map that was used as input for the subregional allocation models.

In addition, SANDAG derived a site-level base year employment inventory from data obtained from a private vendor, National Decision Systems, along with in-house data from previous inventories, telephone directories, and RideLink Car-pool matching records. These records were address-matched for assignment to individual MGRAs.

The forecasting model, as originally run, was unable to allocate regionwide residential growth to the year 2020, because the land base in the model exhausted the regional supply of land planned for urban residential development. A subsequent forecast revision effectively extended the modeled capacity of the 2020 land supply by changing assumptions regarding expected residential densities and development patterns, especially around light and commuter rail station areas and major transit centers. The revised assumptions included the following:

- Full build-out to planned residential densities (outside transit focus areas)
- Higher allowed residential and employment densities within walking distance of transit focus areas to encourage mixed-use development
- Inclusion of residential development within large employment areas

1990 and 1995 Land Use and Land Ownership Inventories   As noted earlier, SANDAG has conducted inventories of land use since 1971. Inventories of land ownership have been compiled since the early 1980s. The regional GIS was used in recent efforts to integrate these two databases within a combined inventory. The land use polygons, generated from interpretations of remote imagery, were aligned to the SanGIS "landbase," composed of lots and parcels. The parcel units were then assigned land use designations from the overlay. This improved the spatial accuracy of the data and allowed consistent updates through time between existing layers. Comparison with other digital data (orthophotographs and street coverages) and input from local jurisdictions enhanced the accuracy of the project. Cross-referencing of parcel designations to the land use codes assigned by the county tax assessor provided further verification of the land use designations.

The development of the land ownership database also involved overlays performed with the SanGIS land base. Data from multiple public agencies (primarily in GIS form or linked to specific parcel records) were used to identify lands in various categories of ownership, including federal, state, and local governments; utility and port districts; public school districts; and others that were distinct from privately held lands.

The resulting inventories were used by SANDAG to track and map regional and subregional trends in activity patterns and the utilization of land resources. Recent land use and ownership type coding schemes represent an increase in detail and complexity of data, allowing for more fine-grained analyses (such as for transportation modeling).

Contacts

Steve Kunkel, Senior Regional Planner

Jeff Tayman, Senior Regional Planner

San Diego Association of Governments: Research and Sources Information Services Division

Sources

San Diego Association of Governments. 1997. *SANDAG Info: Land use in the San Diego region.* July–August.

――――. 1998. *2020 Cities/county forecast—Land use inputs.*

――――. *2020 Cities/county forecast: Overview. Vol. 1.*

――――. *Regional growth management strategy.* <http://www.sandag.cog.ca.us/ftp/html/landuse/regional_growth.html> (May 7, 1998).

――――. *About SANDAG's GIS.* <http://www.sandag.cog.ca.us/ftp/html/projects/ris/gis/aboutgis.html> (October 2, 1998).

SanGIS. Official web site. <http://www.sangis.org> (May 7, 1998).

CASE SUMMARY: LANE COUNCIL OF GOVERNMENTS, OREGON

Overview

The Lane Council of Governments (LCOG) is the metropolitan planning organization for the Eugene-Springfield, Oregon, metropolitan area. LCOG is responsible for administering the Eugene-Springfield urban growth boundary (UGB), which has been in place for more than 15 years. Metro Plan, the regional comprehensive plan, has been undergoing a periodic review, including reassessments of UGB land supply and demand. Furthermore, the update of the regional transportation plan, TransPlan, requires LCOG to initiate a coordinated land use and transportation strategy focusing on some 60 "nodal development areas." Small-scale land supply analyses supported this initiative by assessing the potential for intensive mixed-use development appropriate for other modes of transportation as alternatives to single-occupancy vehicles.

Geographic Information System Resources

The Regional Information System is a multipurpose geographic information system (GIS) maintained by the Lane Council of Governments. The parcel layer (1992 Eugene-Springfield Metro Area Parcel File) is a central component of this system. The currency of the parcel base map varies considerably for different areas of the region served by LCOG; urban areas are generally more up-to-date than the more rural parts of Lane County. Parcel attribute data are derived from assessor's files, permit data (a primary source of information for tracking land use change and for updating street address files), windshield surveys, and aerial photography.

LCOG maintains a vacant developable lands coverage, which is updated annually from aerial photography. The delineation of lands classified as vacant closely follows parcel boundaries in the more urbanized areas, and splits some of the larger outlying parcels to account for vacant portions of partially developed properties.

Other GIS layers include aerial photos, planned land use, floodways, soils, sensitive land areas, slopes, wetlands (national and local surveys), parcel files (tax lots and land use parcels), and site addresses (point coverage).

LCOG has tracked some of its land supply data over time. However, several factors have hindered the utilization of these data for trend analysis. Such hindrances include primarily data management problems such as changes in land use codes over time and the need to convert older data formats into ArcInfo coverages for comparison with more recent mapped data.

Land Supply and Capacity Monitoring

Residential Land and Housing Study (1998)   The *Residential and Housing Study,* undertaken in 1998, represented an update of residential land supply and demand estimates, and reflected major policies related to land supply, which include the following:

- Urban growth areas (UGAs) must contain sufficient lands for 20 years of growth.
- LCOG must endeavor to maintain a 5-year supply of serviced buildable land.
- If the land supply at any time drops below the forecast 20-year demand, LCOG or local jurisdictions must take measures to bring supply and demand into balance.
- Local jurisdictions must increase opportunities for infill and redevelopment.

Several general assumptions guided the land supply and capacity analysis:

- Lands affected by various environmental conditions (see the following lists) were excluded from the inventory.

- Some of the demand would be accommodated by development of low-density lots, infill, and redevelopment (discussed later in this section).
- Thirty-two percent of the gross vacant land supply was deducted for nonresidential uses (e.g., churches, day care centers, parks, and streets).
- Expected future housing-type split was 40 percent single-family detached, 12 percent single-family attached, 35 percent multifamily, and 13 percent manufactured housing.
- The rate of infill observed over the past ten years would continue.

Density assumptions guiding translation of household increase into land demand in acres were derived from the Metro Plan, with adjustments made to single-family detached (lowered) and multifamily (raised) uses to account for observed development trends in low-density areas within each category.

Residential development was planned to occur on lands designated as low-density, medium-density, or high-density. All land with a slope greater than 25 percent, as well as most unserviced land, was classified as low-density. Using information on past development trends, LCOG established average anticipated net densities over the 20-year planning horizon for each housing type and planned density category.

LCOG updated the development status of large lots within each density category using aerial photographs. This produced a refined estimate of buildable residential lands that included both vacant and partially developed parcels. Development status was further updated using recorded approvals for subdivisions and construction covering the period (1992–1995) between the last land use survey and the supply analysis.

For the purposes of the analysis, residential structures were classified as single-family, multifamily, manufactured home in a lot, or manufactured home in a park. (The capacity of manufactured housing parks was estimated through personal contacts with park managers.) LCOG performed a comparative analysis of mix of types and trends (1977–1992) by planning subarea.

LCOG used the 1992 Eugene-Springfield Metro Area Parcel File to calculate aggregate planned residential land use. Unbuildable land, which was subtracted from the inventory, included the following:

- Floodways
- Wetlands (varying criteria, depending on jurisdiction)
- Power line easements
- Land within riparian and water body buffers

Constraints to development, affecting densities and development costs, but not prohibitive to all types of development, included the following:

- Floodplains
- Wetlands (less restrictive criteria than for unbuildable lands)

- Hydric soils
- Slopes greater than 25 percent

The analysis of the undeveloped residential land supply had three distinct components: (1) small parcel land supply, (2) low-density land matrix, and (3) site inventory (as discussed in the following paragraphs).

The site inventory and the low-density land matrix identified "land on which public services were not available and [estimated] the number of years it would take to get services" (primarily sewers). All of the land within the urban growth boundary (UGB) was expected to be serviced within the 20-year planning period.

LCOG summed the buildable land totals from these three analyses to produce a "combined inventory." This inventory did not include land with infill or redevelopment potential. In addition, much of the supply was "constrained" to a development potential below that allowed by regulations.

The Land in Small Parcels component of the study entailed an inventory of all vacant or underdeveloped parcels planned for low-density residential (less than 5 acres) or medium- and high-density residential (less than 1 acre) use, as well as overlay analyses performed to screen out unbuildable land (following the aforementioned criteria). For the purposes of the analysis, "buildable" lots were defined as between 0.1 and 0.25 acre in size and were treated as individual residential lots with capacity for one unit of single-family development.

The Low Density Land Matrix component included an inventory of undeveloped whole or partial tax lots between 5 and 10 acres in size, as well as a polygon overlay to screen out unbuildable lands (following the aforementioned criteria).

The Site Inventory component was an update of a 1991 survey. A "site" was defined as a set of contiguous parcels having the same ownership and planned land use designation. Using the 1992 Metro Area Parcel File, parcels were selected that were either (1) wholly or partially undeveloped lots or (2) parcels with residential zoning or with plan designation of low density, with an area of 10 acres or greater, *or* plan designation of medium or high density with an area of 1 acre or greater.

The three inventories thus generated were compared with the old site inventory to identify discrepancies and needed updates. Several major findings emerged from the combined analysis:

- Demand for housing was expected to continue to shift toward older, smaller, less affluent households.
- Land supply exceeded the full range of demand forecasts for the metropolitan area.
- Metro Plan densities were not being achieved, although the overall net density of new development had been rising.

- Most of the buildable land was near the outer edge of the UGB, although much of that was "constrained" and 28 percent was not serviced with urban infrastructure.
- Residential infill was occurring primarily in areas with large single-family lots and on isolated vacant lots.
- Redevelopment was occurring mainly in downtown areas as multifamily residential replaced parking and other low-intensity land uses.

Evaluating Redevelopment Potential (1997) *Evaluating Redevelopment Potential* was an effort by LCOG to develop a GIS approach for determining redevelopment potential throughout the region. Work in this area focused on both metrowide redevelopment and potential within small areas (nodes). This represented a major addition to previous land supply assessments, as neither redevelopment nor infill had previously been incorporated into this work by LCOG. For the purposes of analysis, LCOG defined redevelopment as "any increase in the intensity of use of a site whether or not it involves alteration of an existing structure." The results of the analysis showed significant redevelopment potential, especially along transportation corridors. This finding informed a "nodal development strategy" for encouraging new development in concentrated areas. Longitudinal data on development activity also showed that significant redevelopment had occurred in some areas.

Analysis of Past Redevelopment Activity. Indicators of past redevelopment were derived from biannual employment data from 1978 to the early 1990s (obtained from the Oregon Employment Division), which were tracked by address. Notably, permit data were deemed not specific enough, and land use records too inconsistent in land use coding over time, to be usable for such longitudinal analysis.

Employment increases (aggregated by census tract), along with anecdotal evidence about significant development projects completed during the study period, led LCOG to conclude that significant redevelopment had occurred, particularly for nonresidential land uses in the downtown areas of Eugene and Springfield.

Proximity Analysis. A proximity analysis utilized a raster GIS technique to identify parcels with potential for redevelopment as indicated by low improvement value per acre relative to other nearby parcels. This analysis was performed for the downtown areas of both Eugene and Springfield.

Parcel data for the areas of study were converted to a 20-foot grid (based on the assumption that each standard-size parcel would likely contain at least one grid cell). Each cell was coded for the improvement value per acre of the parcel within which the grid cell center lay. A 500-foot radius from the center of each cell was used to select proximate cells for comparison. The resulting "neighborhood" computation yielded cells coded with a ratio representing the improvement value per acre of that parcel divided by the average for all of the cells within the analysis radius. Parcels with ratios of 0.2 to 0.4 were assumed to have the greatest potential for redevelopment.

The report identified several weaknesses of the proximity analysis trial. First, the analysis should not have included public lands (such as parks) or parking lots, especially those serving adjacent land uses. These properties, although essentially developed, would have very low assessed valuation for improvements. Second, the analysis did not account for zoning as a variable, especially to the degree that it functioned as a determinant of value and value potential for individual parcels.

In addition, for both computational and methodological reasons, future application of such proximity analysis at the metrowide scale may adopt different neighborhood radii and cell sizes.

Regional Inventory of Sites with Redevelopment and Infill Potential. For a metrowide analysis of redevelopment potential, LCOG focused on the identification of parcels based on mixed criteria, including assessed-land-to-improvement values, lot sizes, current uses, and plan designation.

For residential redevelopment, specific criteria included the following:

- Current use is single-family, duplex, or manufactured dwelling.
- Planned land use is medium- or high-density residential, or mixed-use.
- Improvement value < land value.
- Improvement value per acre < or = $100,000.

For commercial redevelopment, specific criteria included the following:

- Current use is not vacant and not a parking lot (inadequate data were available to identify stand-alone parking lots).
- Planned land use is commercial or mixed-use.
- *Either* improvement value < or = land value *or* improvement value/acre < or = $100,000.

For industrial redevelopment, specific criteria included the following:

- Current use is not vacant and not a parking lot.
- Planned for industrial or mixed use.
- Improvement value < land value.

For residential infill, specific criteria included the following:

- Current use is single-family.
- Built prior to 1970.

- Planned for single-family residential use.
- Parcel is larger than ⅓ acre.
- Slope of 25 percent or less.
- Improvement value per acre < or = $150,000.

(Although the subject was not explicitly addressed in the analysis, the report suggests that land use code can be used to exclude parcels that have possible contamination (brownfields)—a factor that would likely constrain future development.)

The conclusions of the subanalyses included the following:

- For residential infill, the assumption used by LCOG that 32 infill lots will be developed each year was probably too conservative.
- Residential redevelopment findings supported the application of the most restrictive indicator criteria. In the future, areawide proximity analyses should be applied as a refinement to land supply and capacity estimates, largely to account for local variations in land and improvement values.
- The greatest amount of land identified as redevelopable was in the commercial use category. For commercial redevelopment, site-specific barriers were significant, however, suggesting that the estimate may have been overgenerous. Further study to refine the GIS methodology might usefully include case studies of successful redevelopment projects. Furthermore, future analyses should avoid using redundant indicators, such as traffic volume and land value.
- Site-specific barriers were also common in the industrial redevelopment category and difficult to identify with GIS. Ownership was important as well. Owners sometimes hold onto vacant or underutilized land for future expansion. The improvement-to-land-value ratio may be least appropriate for this category, because industrial improvement values are often very low. Finally, identifying lands with potential contamination may be the most fruitful use of GIS in analyzing industrial redevelopment potential (although such work should be done with caution).
- The report qualified the overall findings with an important caveat: "There are many factors that affect whether a parcel will be redeveloped or not over a 20-year period, many of which cannot be analyzed with a GIS; the analysis that can be done with GIS is only a 'first cut' at identifying the actual capacity for this kind of development areawide."

Evaluation of Proposed Nodal Development Areas. The sites identified through the metrowide redevelopment analysis were subsequently summarized for each proposed nodal development area. Each parcel within the nodes was assigned a "redevelopment code," which was based on multiple criteria. Summary findings and subtotals for each node addressed primarily residential and commercial redevelopment potential. Parcel size emerged as an overarching issue. Most vacant lots within the nodal areas were found to be small, posing significant limitations to the redevelopment or infill that could occur on these lands.

Considerations for Future Refinements to the Methodology. Work on this most recent study suggested future refinements to an analysis model for assessing potential for redevelopment in Eugene-Springfield:

- Utilize traffic counts, or intersection analysis, as a basis for evaluating value and use potential for commercial activites.
- Identify independently owned parking lots as a separate category to be included in the inventory of potentially redevelopable land.
- Consider the impact of shared parking, especially church lots adjacent to other uses.
- Consider overhead utilities as barriers to development (although difficult to incorporate into GIS).
- Attempt to analyze the impacts of subregional real estate markets and demographic variables on redevelopment and infill potential.
- Identify potentially contaminated sites (e.g., storage tanks, service stations, and dry cleaners).
- Compare costs of developing different types of projects, such as infill for multifamily versus commercial uses, or redevelopment versus development of vacant land.
- Consider the site orientations of buildings within parcels and the impact of this factor on the space available for additional structures.
- Exclude publicly owned land from most analyses.

Contacts

Cress Bates, GIS Program Manager
Clair Van Bloem, Senior Research Analyst
Lane Council of Governments

Sources

Criterion Engineers/Planners. 1996. *Integration of transportation and land-use efficiency with growth management.* Report prepared for Lane Council of Governments. Lane Council of Governments. 1998. *TransPlan.* Draft.

———. 1998. *Eugene-Springfield metropolitan area: Residential land and housing study: Citizen advisory committee policy recommendations report.*

———. 1997. *Eugene-Springfield metropolitan area residential lands and housing study: Draft supply and demand technical analysis.*

———. 1997. *Evaluating redevelopment potential in the Eugene-Springfield metropolitan area.*

Appendix C

Interview Contacts

The following persons were interviewed during the course of the project, in addition to the professional staff interviewed to develop the case summaries.

Scott Bollens
University of California at Irvine, Department of Urban and Regional Planning

Gilbert Castle
Castle Consulting, San Francisco, California

Ed Crane
Environmental Systems Research Institute, Redlands, California

John De Grove
Florida Atlantic University, Joint Center for Environmental and Urban Problems

Gerald Dildine
Intergraph Corporation, Huntsville, Alabama

William Drummond
Georgia Institute of Technology, City Planning Program

Kenneth Dueker
Portland State University, Center for Urban Studies

Pierce Eichelberger
Urban and Regional Information Systems Association, Park Ridge, Illinois

Joseph Ferreira
Massachusetts Institute of Technology, Department of Urban Studies and Planning

David Godschalk
University of North Carolina at Chapel Hill, Department of City and Regional Planning

William Huxhold
University of Wisconsin-Milwaukee, Department of Urban Planning

Sanjay Jeer
American Planning Association, Research Division, Chicago, Illinois

Richard Klosterman
University of Akron, Department of Geography and Planning

Gerrit Knaap
University of Illinois at Urbana-Champaign, Department of Urban and Regional Planning

John Landis
University of California at Berkeley, Department of City and Regional Planning

Bob Lima
Bosch Institute, Hyannis, Massachusetts

Stewart Meck
American Planning Association, Research Division, Chicago, Illinois

Terry Moore
ECO Northwest, Portland, Oregon

Zorica Nedovic-Budic
University of Illinois at Urbana-Champaign, Department of Urban and Regional Planning

Arthur C. Nelson
Georgia Institute of Technology, City Planning Program

Bernard Niemann
University of Wisconsin-Madison, Department of Urban and Regional Planning

Gary Pivo
University of Washington, Department of Urban Design and Planning

Douglas Porter
Growth Management Institute, Chevy Chase, Maryland

George Rolfe
University of Washington, Department of Urban Design and Planning

Grant Thrall
University of Florida, Geography Department

Mary Tsui
California Geographic Information Association, Sacramento, California
Land Systems Group, Monterey, California

Nancy von Meyer
Fairview Industries, Blue Mounds, Wisconsin

Ric Vrana
Portland State University, Department of Geography

Paul Waddell
University of Washington, Department of Urban Design and Planning, Evans
School of Public Affairs

Appendix D

May 1998
Seminar Participants

The following persons participated in a seminar entitled "Parcel-Based GIS for Land Supply and Capacity Monitoring" and held at the University of Washington, May 1998.

Marina Alberti
University of Washington, Department of Urban Design and Planning

Carol Bason
City of Seattle, Washington

Earl Bell
University of Washington, Department of Urban Design and Planning

Bill Beyers
University of Washington, Department of Geography

Eric Bishop
Environmental Systems Research Institute, Olympia, Washington

Glenn Bolen
Portland Metro, Oregon

Monitoring Land Supply with Geographic Information Systems, edited by Anne Vernez Moudon and Michael Hubner ISBN 0 471371673 © 2000 John Wiley & Sons, Inc.

Scott Bollens
University of California at Irvine, Department of Urban and Regional Planning

Jennifer Bradford
Portland Metro, Oregon

Nicholas Chrisman
University of Washington, Department of Geography

Keith Dearborn
Attorney at Law, Seattle, Washington

Ken Dueker
Portland State University, Center for Urban Studies

Miles Erickson
University of Washington, Center for Community Development and Real Estate

Chandler Felt
King County, Washington

Bob Filley
University of Washington, Center for Community Development and Real Estate

David Godschalk
University of North Carolina at Chapel Hill, Department of City and Regional Planning

Aaron Hoard
City of Bellevue, Washington

Shane Hope
Washington State Department of Community, Trade, and Economic Development

Lewis Hopkins
University of Illinois at Urbana-Champaign, Department of Urban and Regional Planning

George Horning
King County, Washington

Michael Hubner
University of Washington, Cascadia Community and Environment Institute

William Klein
American Planning Association, Chicago, Illinois

Gerrit Knaap
University of Illinois at Urbana-Champaign, Department of Urban and Regional Planning

Donald Miller
University of Washington, Department of Urban Design and Planning

Anne Vernez Moudon
University of Washington, Department of Urban Design and Planning

Peter Moulton
Washington State Department of Community, Trade, and Economic Development

Lori Peckol
Puget Sound Regional Council

Peter Robinson
Intergraph Corporation, Kirkland, Washington

George Rolfe
University of Washington, Department of Urban Design and Planning

Nancy Tosta
Puget Sound Regional Council

Stephen Toy
Snohomish County, Washington

Glenda Van Engelen
Intergraph Corporation, Kirkland, Washington

Ric Vrana
Portland State University, Department of Geography

Paul Waddell
University of Washington, Department of Urban Design and Planning, Evans School of Public Affairs

Frank Westerlund
University of Washington, Department of Urban Design and Planning

Glossary of Terms and Acronyms

Adjusted supply/capacity   Reduction of maximum supply/capacity (see definition) to account for factors that mitigate against full build-out, including land market behavior, land development and construction practices, and the provision of public services. Also referred to as "available" or "net" supply or capacity.

Baseline land inventory   An inventory of current (or recent) land use, supply, or capacity against which to measure future changes. Often contains not only information on stocks, but also rates of change or activity, including recent trends in densities and land use mix achieved, and well as timing and location of development.

Brownfields development   Development of abandoned, idled, or underused industrial and commercial sites where expansion or redevelopment is complicated by real or perceived environmental contamination (Natural Resources Defense Council 1999).

Buildable land supply   Land within the overall supply on which additional or new development can occur within regulatory, physical, or market-imposed limits. Expressed as an amount of land, such as acres.

Cadastre   Public record, survey, or map of the value, extent, and ownership of land as a basis of taxation (*Websters Second New Riverside University Dictionary* 1988).

Demand allocation   The conversion of regional forecasts of population and employment into subregional forecasts or "targets" for growth over a speci-

For source references in this glossary, see Appendix F, "General Bibliography"

fied time period. Considerations of local land supply or capacity limits may or may not be included to adjust this allocation.

Development capacity   The amount of additional or new development that can occur on land identified as "buildable." Expressed as an amount of development, such as dwelling units or building square footage.

Development constraint   Any of various physical, environmental, social, or market-related factors that function to limit the potential of the land supply to accommodate new or additional development.

Geographic information system (GIS)   1. Computerized database system for capture, storage, retrieval, analysis, and display of spatial data (Huxhold 1991). 2. System of hardware, software, and procedures designed to support the capture, management, manipulation, analysis, modeling, and display of spatially referenced data for solving complex planning and management problems (NCGIA 1990, in Heikkila 1998). 3. The organized activity by which people *measure* aspects of geographic phenomena and processes; *represent* these measurements, usually in the form of a computer database, to emphasize spatial themes, entities, and relationships; *operate* upon these representations to produce more measurements and to discover new relationships by integrating disparate sources; and *transform* these representations to conform to other frameworks of entities and relationships (Chrisman 1997, emphasis in original).

Greenfields development   The conversion of land from nonurban to urban uses, primarily occurring on large parcels (or groups of parcels) at the fringes of urbanized regions.

Gross density   Quantitative measure of development (e.g., housing units, building square footage) per land area to include private lands and related public lands—multiple parcels, rights-of-way, and other areas of land not contained in parcels containing this development. Land areas considered can be defined by jurisdictional boundaries or by other boundaries. Gross density can also be established by land use category. (See also "net density" definition.)

Infill development   Development of vacant lands that lie within areas that have largely been developed for urban uses. Both in the literature and in practice, infill development is neither clearly nor consistently distinguished from vacant land development or redevelopment. Within the practice of land supply and capacity monitoring (LSCM) "infill" is often operationally defined in relation to partially utilized parcels that may be subdivided to accommodate new development. (*See also* Partially utilized land.)

Land capacity   Total amount of development (may include existing development) that can occur on a given supply of land. Usually defined as number of dwelling units or square feet of built nonresidential space.

Land demand   Translation of demographic and economic projections into specific location and space requirements for households, firms, and supporting public services (Kaiser et al. 1995).

Land information system (LIS)   1. System for acquiring, processing, storing, and distributing information about land (Dale and McLaughlin 1988). 2. Geographic information system having, as its main focus, data concerning land records—to include resource, land use, environmental impact, and fiscal data (Dueker and Kjerne 1989).

Land supply   All land within an urbanized region, including developed, underdeveloped, and partially developed land, along with all of the characteristics and conditions affecting the future use of land in each of these categories. Expressed as an amount of land (e.g., acres).

Land use inventory   Collection, classification, enumeration, and mapping of data on urban land as functional space devoted to various uses with reference to specific activities, locations, and structures (Kaiser et al. 1995).

Land use ratio   Breakdown of categories of land use as percentages of the total amount of land in a community. May include public use lands, such as rights-of-way, parks, or schools, as well as categories of residential, commercial, and industrial development. Often used as a basis for projecting gross land requirements (Harris 1992).

Legacy data   Archived longitudinal data related to land use, supply, or capacity.

Longitudinal data   Data captured over multiple points in time, which may be compared to derive amounts and rates of change in land use, supply, or capacity over time.

Market factor   Percent deduction from land supply and capacity estimates to account for land kept off the market because of speculation, land banking, future expansion, personal use, and other related factors.

Maximum supply/capacity   Greatest potential amount of land that can be developed, or development that can occur, within the parameters set by zoning, infrastructure, servicing, and environmental constraints. Also referred to as "theoretical" or "gross" supply or capacity, or simply "build-out."

Metadata   Data describing GIS data, data set, or database, addressing data transfer, lineage, currency, accuracy, extent, custodianship, and collection methodology, as well as applicable data-processing algorithms or procedures (New York State Office of Real Property Services 1996; British Columbia Ministry of Environment, Lands, and Parks 1997).

Multiple Listing Service (MLS)   Local real estate marketing database listing properties currently for sale or recently sold. May include such data as asking price, time on market, detailed descriptions of land and improvements, utility servicing, and zoning.

Multipurpose cadastre   Large-scale land information system that employs the parcel (tax lot) as the fundamental unit of spatial organization and relates a series of land records to this parcel, including land rights, taxation, use, structures, utility servicing, physical conditions, and other data (Dale and McLaughlin 1988).

Neighborhood operation (analysis)   GIS function involving assembly within a specific "neighborhood" (usually defined by distance) on a single spatial layer (Chrisman 1997). Analysis option for parcel data in raster GIS format—used to generate new parcel attributes based on the attributes of other nearby parcels.

Net density   Quantitative measure of development (housing units, square footage) per land area within individual, usually privately held, parcels, often summarized by land use category (e.g., residential, commercial).

Orthophotography   Photograph of the earth's surface in which geographic distortion has been removed (New York State Office of Real Property Services 1996).

Parcel   Physical area of land, with legally defined boundaries, that is owned by one person or entity; also referred to as a lot or a tax lot (Huxhold 1991).

Parcel-based geographic information system (PBGIS)   A GIS that includes a parcel base map as a primary spatial layer, with land records files linked to parcel features via a unique identifier.

Parcel identification number (PIN)   Unique numerical identifier assigned to individual parcel records, providing a means to link mapped parcel features in a GIS with parcel attribute files.

Partially utilized land   Land (usually defined in terms of parcels) that is vacant in significant part and that allows for development in addition to that existing. (*See also* Infill development.)

Permit tracking system   Automated system for tracking building and subdivision permit data, potentially providing for a transaction-based update of parcel records via common PIN or standardized address identification (Vrana and Dueker 1996).

Pipeline development   Projects in the process of being reviewed for public action—including zoning, subdivision, building, or special permits—as well as projects approved for development but unbuilt (Godschalk et al. 1986; Kaiser et al. 1995).

Planning support system (PSS)   Computer hardware, software, and related information that are used for planning, including support for continuous and interactive analysis, design, and evaluation activities, and provision of structured and accessible information for executive and public decision making (Klosterman 1997).

Parcel points   GIS layer representing parcels as geocoded locations, which correspond either to parcel centroids (xy coordinates) or to address ranges along a street network.

Polygon overlay Spatial analysis function that uses Boolean logic (AND, OR, NOT, etc.) to create new polygons from the intersection of the boundaries of polygons from two or more GIS layers (Huxhold 1991).

Potential projected supply/capacity Assessment of land, its regulation, and potential for future development. May include evaluation of future policy alternatives, as well as efforts to account explicitly for variable time frames and changes in relevant factors over time (e.g., changes in zoning, land market conditions, etc.).

Raster Spatial data model based on a rectangular division of a geographic surface into pixels or grid cells that are coded with measured attribute values (Chrisman 1997).

Redevelopment Demolition and replacement, or significant alteration of structures on a site to accommodate more intense, highly valued, or different land uses.

Regulatory envelope Maximum amount of development (defined by physical measures, such as floor area ratio, building coverage, setback, and height restrictions) allowed by zoning and other development regulations. Also called zoning envelope.

Serviced land Land provided with current or planned urban levels of infrastructure and urban services, including (primarily) transportation, water, and sewer and (secondarily) schools, parks, fire and police protection, telecommunications, and other services.

Spatial analysis Generation of new information based on explicit spatial processing of features on either single or multiple layers of GIS data (Webster 1993).

Suitability analysis Measure of the physical capacity and appropriateness of a site to accommodate specific land uses, including considerations of environmental and infrastructure capacity, social acceptability, and economic feasibility (Hopkins 1977; Anderson 1987).

Tax lot *See* parcel.

Transportation/forecast analysis zone (TAZ/FAZ) Delineated subareas of an urban region utilized for the collection of activity data, as well as for the allocation of residential and employment growth (Kaiser et al. 1995).

Under-build Land developed at densities that are significantly lower than those allowed by subdivision, zoning, and other development regulations (includes both residential and nonresidential land uses).

Underutilized land Land (usually defined in terms of parcels) that is not vacant and zoned for more intensive uses than those that occupy it. Also land with improvements that are of relative low value. (See "redevelopment" definition.)

Unit of analysis Geographic unit for which land supply and/or capacity is estimated in accordance with a set of analytical assumptions and operations.

Unit of data collection Geographic unit for which data related to land supply and capacity are captured and entered into a database.

Vacant land Land lacking urban use or significant improvements.

Vector Spatial data model based on the geometry of points, lines, and areas and located by coordinate measurements in a spatial reference system (Chrisman 1997).

General Bibliography

American Planning Association. 1998. *Growing Smart legislative guidebook*. Phase II. Interim ed. Chicago: American Planning Association.

Anderson, Larz T. 1987. *Seven methods for calculating land capability/suitability*. Planning Advisory Service Report Number 402. Chicago: American Planning Association.

Bain, B., and S. Jeer. 1997. *LBCS annotated bibliography*.
⟨http://www.planning.org/plnginfo/lbcs/index.html⟩ (December 13, 1997).

Bechhoefer, Ina S. 1996. Entering the information age: Transforming real estate data into meaningful information will redefine organizational and client communication. *Urban Land* 3:37–39, 57–60.

Birkin, Mark, Graham Clarke, Martin Clarke, and Alan Wilson. 1996. *Intelligent GIS: Location decisions and strategic planning*. Cambridge, U.K.: GeoInformation International; New York: John Wiley & Sons.

Bollens, Scott A. 1998. Land supply monitoring systems. In *The Growing Smart working papers*. Vol. 2. Chicago: American Planning Association.

Bollens, Scott A., and David R. Godschalk. 1987. Tracking land supply for growth management. *Journal of the American Planning Association* 3:315–327.

British Columbia Ministry of Environment, Lands, and Parks. *GIS glossary page*. December 18, 1997. ⟨http://www.env.gov.bc.ca/gis/glosstxt.html⟩ (June 1, 1999).

Brower, David J., David R. Godschalk, and Douglas R. Porter, eds. 1989. *Understanding growth management: Critical issues and a research agenda*. Washington D.C.: Urban Land Institute.

Castle, Gilbert H. 1997a. Location, location, location–Not! *Business Geographics* 4:18.

———. 1997b. The bigger picture. *Business Geographics* 2:16.

————, ed. 1993. *Profiting from a geographic information system*. Fort Collins, Colo.: GIS World Books.

Charlton, Martin, and Simon Ellis. 1991. GIS in planning. *Planning Outlook* 1:20–26.

Chrisman, Nicholas. 1997. *Exploring geographic information systems*. New York: John Wiley & Sons.

Dale, Peter F., and John D. McLaughlin. 1988. *Land information management: An introduction with special reference to cadastral problems in third world countries*. Oxford: Clarendon Press.

Dearborn, Keith W., and Ann M. Gyri. 1993. Planner's panacea or Pandora's box: A realistic assessment of the role of urban growth areas in achieving growth management goals. *University of Puget Sound Law Review* 3:975–1024.

De Grove, John M. 1992. *Planning and growth management in the states*. Cambridge, Mass.: Lincoln Institute of Land Policy.

Diamond, Henry L., and Patrick F. Noonan, eds. 1996. *Land use in America*. Washington D.C.: Island Press.

Ding, Chengri, Lewis Hopkins, and Gerrit Knaap. 1997. Does planning matter? Visual examination of urban development events. *Landlines* 1:4–5.

Donahue, James G. 1994. Cadastral mapping for GIS/LIS. *ACSM/ASPRS international proceedings page*. ⟨http://wwwsgi.ursus.maine.edu/gisweb/spatdb/acsm/ac94114.html⟩ (January 22, 1999).

Downs, Anthony. 1994. *New visions for metropolitan America*. Washington, D.C.: Brookings Institute; Cambridge, Mass.: Lincoln Institute of Land Policy.

Dueker, Kenneth J., and P. Barton DeLacey. 1990. GIS in the land development planning process: Balancing the needs of land use planners and real estate developers. *Journal of the American Planning Association* 4:483–491.

Dueker, K. J., and Daniel Kjerne. 1989. *Multipurpose cadastre: Terms and definitions*. Falls Church, Va.: American Society for Photogrametry and Remote Sensing and American Congress on Surveying and Mapping.

Easley, V. Gail. 1992. *Staying inside the lines: Urban growth boundaries*. Planning Advisory Service Report Number 440. Chicago: American Planning Association.

Enger, Susan C. 1992. *Issues in designating urban growth areas*. Parts I and II. Olympia, Wash.: State of Washington, Department of Community Development, Growth Management Division.

Federal Geographic Data Committee, Cadastral Subcommittee. 1996. *Cadastral data content standard for the national spatial data infrastructure*. ⟨http://www.fgdc.gov/standards/documents/standards/cadastral/cadstandard.pdf⟩ (October 10, 1997).

Fernandez-Falcon, Eduardo, James R. Strittholt, Abdulaziz I. Alobaida, Robert W. Schmidley, John D. Bossler, and J. Raul Ramirez. 1993. A review of digital geographic information standards for the state/local/user. *URISA Journal* 2:21–27.

Gemini Consulting. *C4 Lab glossary of terms: Essential guide to convergency terminology page*. ⟨http://digital.gemconsult.com/glossary/index.htm⟩ (January 22, 1999).

Godschalk, David R., Scott A. Bollens, John S. Hekman, and Mike E. Miles. 1986. *Land supply monitoring: A guide for improving public and private urban development decisions*. Boston: Oelgeschlager, Gunn & Hain, in association with the Lincoln Institute of Land Policy.

Harris, Christopher. 1992. Bringing land use ratios into the '90s. *PAS Memo* 8:1–4.

Heikkila, Eric J. 1998. GIS is dead; Long live GIS! *Journal of the American Planning Association* 3:350–360.

Heilbrun, James, with the assistance of Patrick A. McGuire. 1987. *Urban economics and public policy*. New York: St. Martin's Press.

Holmberg, S. C. 1994. Geoinformatics for urban and regional planning. *Environment and Planning B: Planning and Design* 21:5–19.

Hopkin, Lewis D. 1977. Methods for generating land suitability maps: A comparative evaluation. *Journal of the American Institute of Planners* 4:386–400.

Hubner, Michael Henry. 1999. *Urban land supply monitoring: A critical review of current practices in measurement, analysis, and application*. Master's thesis, University of Washington.

Huxhold, William E. 1991. *An introduction to urban geographic information systems*. New York: Oxford University Press.

Innes, Judith E., and David M. Simpson. 1993. Implementing GIS for planning: Lessons from the history of technological innovation. *Journal of the American Planning Association* 2:230–236.

Jacobs, H. M. 1988. *Land information systems and land use planning: An annotated bibliography of social, political and institutional issues*. CPL Bibliography Series. No. 208. Chicago: Council of Planning Librarians.

Johnson, Alan George. 1991. *Three applications of residential capacity analysis in Renton, Washington*. Master's thesis, University of Washington.

Kaiser, Edward J., David R. Godschalk, and F. Stuart Chapin. 1995. *Urban land use planning*. Urbana and Chicago: University of Illinois Press.

Kivell, Philip. 1993. *Land and the city: Patterns and processes of urban change*. London and New York: Routledge.

Klosterman, Richard. 1997. Planning support systems: A new perspective on computer-aided planning. *Journal of Planning Education and Research* 17:45–54.

Knaap, Gerrit, and Arthur C. Nelson. 1992. *The regulated landscape: Lessons on state land use planning from Oregon*. Cambridge, Mass.: Lincoln Institute of Land Policy.

Knaap, Gerrit, Lewis Hopkins, and Arun Pant. 1996. Does transportation planning matter? A framework for examining the logic and effects of planned transportation infrastructure on real estate sales, land values, building permits, and development sequence. *DPM: Does planning matter? page*. ⟨http://www.urban.uiuc.edu/projects/portland/lincoln.html⟩ (November 1, 1997).

Kollin, Cheryl, Lisa Warnecke, Winifred Lyday, and Jeff Beattle. 1998. Growth surge: Nationwide survey reveals GIS soaring in local governments. *GeoInfo Systems* 2:25–30.

Korte, George. 1997. *The GIS book*. Santa Fe: OnWord Press.

Landis, John D. 1994a. The California urban futures model: A new generation of metropolitan simulation models. *Environment and Planning. B*. 4:399–420.

———. 1994b. Future tense. *Planning* 2:22–25.

Landis, John, and Paul Sedway. 1996. A quarter-century of environmental regulation and growth control in California. *Urban Land* 10:89–92.

Mildner, Gerard C. S., Kenneth J. Dueker, and Anthony M. Rufolo. 1996. *Impact of the urban growth boundary on metropolitan housing markets*. Portland, Oreg.: Portland State University for Urban Studies.

Miles, Mike E., Richard L. Haney Jr., and Gayle Berens. 1996. *Real estate development: Principles and process.* Washington D.C.: Urban Land Institute.

Moore, C. A., C. F. Donaldson, and R. C. Burrus. 1995. GIS supports urban rezoning. *GIS World* 2:61–63.

Morris, Marya. 1996. *Creating transit-supportive land-use regulations.* PAS Report No. 468. Chicago: American Planning Association.

Natural Resources Defense Council. 1999. *Lookup page.* ⟨http://www.nrdc.org/sitings/fslook.html⟩ (January 25, 1999).

Nedovic-Budic, Zorica, and Jeffry K. Pinto. 1998. Coordinating development and use of geographic information system databases. Unpublished project report, Champaign: University of Illinois.

Newcomb, Tod. 1994. GIS: Not just a pretty map. *Urban Land* 5:21–24.

New York State Office of Real Property Services. 1996. *Statewide GIS coordination plan page.* ⟨http://www.orps.state.ny.us/gis/projects/sgcplan/⟩ (January 22, 1999).

Onsrud, Harlan J., and Gerard Rushton, eds. 1995. *Sharing geographic information.* New Brunswick, N.J.: Center for Urban Policy Research.

Orman, Larry. 1997. Computer mapping: New frontier for land use and real estate professionals. *Lusk Review* 1:51–61.

Peirce, Neil. 1997. Maryland's "Smart Growth" law: A national model? *Washington Post Writers Group page.* ⟨http://www.clearlake.ibm.com/Alliance/newstuff/peirce/peirce_042097.html⟩ (October 21, 1997).

Porter, Douglas R. 1995. A 50-year plan for metropolitan Portland. *Urban Land* 7:37–40.

Public Technology, Inc., Urban Consortium, and International City Management Association. 1991. *Local government guide to geographic information systems: Planning and implementation.* Washington D.C.: Public Technology, Inc., Urban Consortium, and International City Management Association.

Real Estate Research Corporation. 1982. *Infill development strategies.* Washington D.C.: Urban Land Institute and American Planning Association.

Rosenberg, D., et al. 1995. *Beyond sprawl: New patterns of growth to fit the new California.* San Francisco: Bank of America with California Resources Agency, Greenbelt Alliance, and Low Income Housing Fund.

Simons, Robert, and Mark Salling. 1995. Using GIS to make parcel-based real estate decisions for local government: A financial and environmental analysis of residential lot redevelopment in a Cleveland neighborhood. *URISA Journal* 1:7–19.

Tulloch, David L., Bernard J. Niemann, Earl F. Epstein, W. Frederick Limp, and Shelby Johnson. 1996. Comparative study of multipurpose land information systems development in Arkansas, Ohio, and Wisconsin. In *GIS/LIS '96: Annual conference and exposition proceedings,* 128–141. Bethesda, Md.: American Society for Photogrammetry and Remote Sensing.

Ventura, Stephen J. 1991. *Implementation of land information systems in local government—Toward land records modernization in Wisconsin*: Madison: Wisconsin State Cartographer's Office.

von Meyer, Nancy, and D. David Moyer. 1994. New wheels for a new highway: Land information standards. *URISA News* 141:1, 6.

Vrana, Ric, and Kenneth J. Dueker. 1996. *LUCAM: Tracking land use compliance and monitoring at Portland Metro*. Report to Metro, Portland, Oregon. Portland: Center for Urban Studies, Portland State University.

Webber, Melvin M. 1965. The roles of intelligence systems in urban-systems planning. *Journal of the American Institute of Planners* 11:289–296.

Webster, C. J. 1993. GIS and the scientific inputs to urban planning: Part 1. Description. *Environment and Planning B: Planning and design*. 20:709–728.

———. 1994. GIS and the scientific inputs to urban planning: Part 2: Prediction and prescription. *Environment and Planning. B. Planning and Design* 21:145–157.

Webster's Second new Riverside university dictionary. 1988. Boston: Riverside Publishing Company.

Index

9 780471 371632